ADVANCES IN PROTEIN CHEMISTRY AND STRUCTURAL BIOLOGY

Volume 76

Structural Genomics, Part B

ADVANCES IN PROTEIN CHEMISTRY AND STRUCTURAL BIOLOGY

SERIES EDITORS

ALEXANDER McPHERSON
University of California, Irvine
Department of Molecular Biology
and Biochemistry
USA

DAVID S. EISENBERG
Department of Chemistry and Biochemistry
Center for Genomics and Proteomics
University of California, Los Angeles
Los Angeles, California

VOLUME 76

Structural Genomics, Part B

EDITED BY

ANDRZEJ JOACHIMIAK
Midwest Center for Structural Genomics and
Structural Biology Center, Biosciences Division,
Argonne National Laboratory, Illinois, USA

AMSTERDAM • BOSTON • HEIDELBERG • LONDON
NEW YORK • OXFORD • PARIS • SAN DIEGO
SAN FRANCISCO • SINGAPORE • SYDNEY • TOKYO
Academic Press is an imprint of Elsevier

Academic Press is an imprint of Elsevier
The Boulevard, Langford Lane, Kidlington, Oxford OX5 1GB, UK
30 Corporate Drive, Suite 400, Burlington, MA 01803, USA
525 B Street, Suite 1900, San Diego, CA 92101-4495, USA

First edition 2009

ISBN: 978-0-12-374442-5
ISSN: 1876-1623

For information on all Academic Press publications visit our website at www.elsevierdirect.com

Printed and bound in USA
09 10 11 12 10 9 8 7 6 5 4 3 2 1

Contents

Preparation and Characterization of Bacterial Protein Complexes for Structural Analysis

Allan Matte, Guennadi Kozlov, Jean-François Trempe, Mark A. Currie, David Burk, Zongchao Jia, Kalle Gehring, Irena Ekiel, Albert M. Berghuis, and Miroslaw Cygler

Strategies for the Cloning and Expression of Membrane Proteins

Christopher M. M. Koth and Jian Payandeh

PREPARATION AND CHARACTERIZATION OF BACTERIAL PROTEIN COMPLEXES FOR STRUCTURAL ANALYSIS

By ALLAN MATTE,* GUENNADI KOZLOV,† JEAN-FRANÇOIS TREMPE,† MARK A. CURRIE,‡ DAVID BURK,† ZONGCHAO JIA,‡ KALLE GEHRING,† IRENA EKIEL,*,§ ALBERT M. BERGHUIS,† AND MIROSLAW CYGLER*,†

*Biotechnology Research Institute, National Research Council Canada, Montreal, Quebec, Canada
†Department of Biochemistry, McGill University, Montreal, Quebec, Canada
‡Department of Biochemistry, Queen's University, Kingston, Ontario, Canada
§Department of Chemistry and Biochemistry, Concordia University, Montreal, Quebec, Canada

ADVANCES IN PROTEIN CHEMISTRY AND STRUCTURAL BIOLOGY, Vol. 76
DOI: 10.1016/S1876-1623(09)76001-2

Abstract

Bacteria mediate a large variety of biological processes using protein complexes. These complexes range from simple binary heterodimeric enzymes to more complex multi-subunit complexes that can be described as macromolecular machines. A key to understanding how these complexes function is obtaining structural information using methods that include electron microscopy, small-angle X-ray scattering, NMR spectroscopy, and X-ray crystallography. Here we describe a variety of approaches to the expression, purification, and biophysical characterization of bacterial protein complexes as a prerequisite to structural analysis. We also give several examples of the kinds of information these different biophysical approaches can provide and various experimental approaches to obtaining structure information for a given system. Further, we describe several examples of protein complexes where we have obtained structural data that have led to new biological insights.

I. Introduction

Cellular processes involve interactions between multiple proteins. There are many well-documented examples of this in microbial systems, including the co-localization of enzymes associated with sequential steps within an enzymatic pathway, sometimes found as a hetero-oligomeric complex (Klem and Davisson, 1993); the association of proteins to effect cell division (Gamba *et al.*, 2009); the assembly of proteins into complexes to create pores or channels in order to import or export molecules through one or more membranes (Collins *et al.*, 2007); and the control of gene expression through the modulation of both transcription (Rutherford *et al.*, 2009) and translation (Yu *et al.*, 2009). These protein–protein interactions span a wide range of timescales and affinities adapted to their specific biological roles.

We can differentiate between relatively weak and transient protein–protein interactions, which exist as part of many cellular functions, versus stronger interactions resulting in longer lived and more stable protein complexes, in which proteins come together to effect function as an unit. On one end of this spectrum are hetero-oligomeric enzymes that contain two or more subunits usually tightly associated together, while on the other end are proteins that exist independently and associate into meta-stable complexes in order to perform a specific task. Structural analysis of protein complexes, using methods such as X-ray crystallography and NMR spectroscopy, offers a way to visualize at the molecular level protein–protein interfaces and analyze their properties. Many important questions in this area need to be addressed. What dictates the specificity of association? How does one protein interact with multiple partners? What are the structural adjustments at protein interfaces upon association? What features of the surface define interaction "hot spots"? What are the important molecular interactions that govern association and dissociation? What role does conformational change play in interaction? Answers to these questions will yield a new level of understanding at the molecular level and provide a basis to modulate the strength of protein association and allow engineering of new interacting proteins that exert their function *in vivo*.

There is a significant and rapidly growing number of protein complexes deposited in the Protein Data Bank (http://www.rcsb.org/pdb/home/home.do). Several derivative databases have been created in recent years to navigate and curate such complexes. Some examples of these databases are PROTCOM (http://www.ces.clemson.edu/compbio/protcom (Kundrotas and Alexov, 2007)), 3D Complex (http://supfam.mrc-lmb.cam.ac.uk/elevy/3dcomplex/Home.cgi (Levy *et al.*, 2006)), and SNAPPI-DB (http://www.compbio.dundee.ac.uk/SNAPPI/snappidb.jsp (Jefferson *et al.*, 2007)). These databases can serve as useful resources for biologists in various disciplines in order to correlate structural information with *in vivo* and *in vitro* data on particular systems of interest. Analysis of such data at the molecular level also helps to shed light on fundamental questions in protein biochemistry, such as the differences between specific and non-specific protein–protein interfaces (Bahadur *et al.*, 2004; Kobe *et al.*, 2008). A number of other databases accumulate experimental information on protein complexes and protein–protein interactions, many of which are listed at http://www.imb-jena.de/jcb/ppi/jcb_ppi_databases.html.

In this chapter we will summarize our experiences regarding the preparation of bacterial protein complexes for structural analysis, with a specific emphasis on protein complexes from the model bacterium *Escherichia coli*. The structural characterization of protein complexes poses special experimental challenges and requires attention to details that to some extent are unique to each system and in some cases different from those for individual proteins.

II. Protein Complexes of Bacteria

Bacteria contain a wealth of both soluble and membrane-bound protein complexes, as is clearly evident from the enormous literature on this topic and numerous databases devoted to cataloging these molecules. These complexes range from those that can truly be described as "molecular machines" such as the ribosome (Steitz, 2008), the RNA degradosome (Carpousis, 2007), and the protein translocation machinery (Driessen and Nouwen, 2008) to much more modest pairs of relatively small proteins. Systematic large-scale studies of bacterial protein interaction networks at the level of whole proteomes have been initiated using the yeast 2-hybrid (Y2H) method as applied to a subset of the genome of *Helicobacter pylori* (Rain *et al.*, 2001), *Synechocystis* sp. PCC6803 (Sato *et al.*, 2007), and *Campylobacter jejuni* (Parrish *et al.*, 2007). The agreement between different methods applied to the same genome is not particularly good, although in a favorable case of 31 proteins from the *H. pylori* type IV secretion system (Terradot *et al.*, 2004) co-purification experiments authenticated 76% of the interactions predicted from large-scale Y2H experiments (Rain *et al.*, 2001). Blue native/SDS-PAGE (sodium dodecyl sulfate polyacrylamide gel electrophoresis) (Wittig and Schagger, 2008) has been used to identify additional protein complexes from *H. pylori* (Lasserre *et al.*, 2006).

As with many aspects relating to bacterial biochemistry and physiology, a wealth of data relating to protein complexes is available for *E. coli*. This data has been generated over many years from a large number of investigator-driven studies, as well as large-scale interactome analysis of the *E. coli* K-12 genome. These large-scale studies include tandem-affinity purification (TAP-tagging (Collins and Choudhary, 2008)) to isolate protein complexes, followed by their identification by mass spectrometry MS–MS analysis (Butland *et al.*, 2005) and pull downs of His-tagged proteins from the ASKA library (Kitagawa *et al.*, 2005), followed by matrix-assisted laser desorption/ionization-time of flight (MALDI-TOF) analysis (Arifuzzaman *et al.*, 2006). Although these two techniques rely on a similar principle, the overlap of bona fide protein complexes between these two independent datasets is rather low, making it difficult to use as a source of information for selecting protein complexes for structural studies. Recent studies on the *E. coli* protein interaction network suggest that essential proteins make up a significant "core" of the experimentally identified interactions (Lin *et al.*, 2009). It has also been shown that proteins that participate in the same interaction share similar mRNA half-lives, a relationship not previously noted (Janga and Babu, 2009). An alternative approach, the use of blue native/SDS-PAGE, followed by the identification of the protein complex by LC–MS/MS, has been used to identify hetero-oligomeric complexes from *E. coli* (Lasserre *et al.*, 2006). Other studies have focused on characterizing another, although more experimentally challenging set of protein complexes, those found in the membrane (Stenberg *et al.*, 2005).

Collectively, the experimentally determined protein–protein interaction data for *E. coli* have been captured within various databases, such as the Bacteriome.org portal (http://www.compsysbio.org/bacteriome/ (Su *et al.*, 2008)) and the microbial protein interaction database, MPIDB (http://www.jcvi.org/mpidb/about.php (Goll *et al.*, 2008), and are available for browsing using a variety of criteria (Su *et al.*, 2008). A number of computational approaches have also been applied to predict protein interactions in *E. coli*, including methods relying on gene fusions (Enright *et al.*, 1999), functional linkages (Yellaboina *et al.*, 2007), and the interacting domain profile pair method (Wojcik *et al.*, 2002). The extent to which these predictive methods correlate with the high-throughput data on physical interactions is yet to be established. A promising new approach for the discovery and validation of protein interaction networks in bacteria is the application of one-step inactivation of chromosomal genes in *E. coli* (Datsenko and Wanner, 2000). This method permits the high-throughput construction of double knockouts that allows analysis of synthetic lethality of gene product pairs (Butland *et al.*, 2008). It is important to keep in mind that physical and genetic interactions are highly complementary. As was shown in yeast, less than 1% of synthetic lethal genetic interactions were also observed physically (Tong *et al.*, 2004). A new and rapidly developing branch of bioinformatics deals with integrating and "cleaning" high-throughput experimental data (Beyer *et al.*, 2007; Liu *et al.*, 2008; Patil and Nakamura, 2005). It is clear from the comparison of results obtained by different methods and

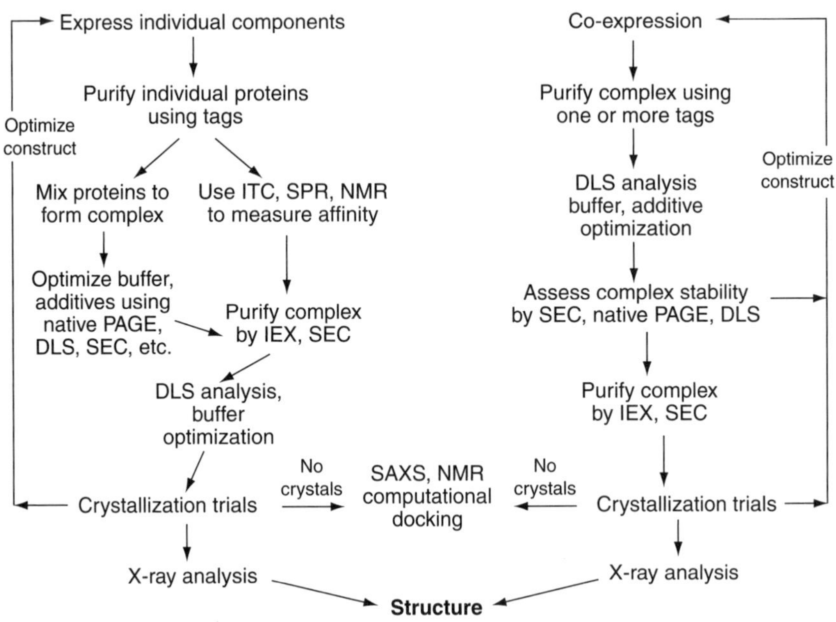

Fig. 1. Summary of experimental strategies used to prepare and characterize a bacterial protein complex.

different investigators that the high-throughput data indicating pair-wise interactions contain a significant fraction of false positives. Therefore, structural investigation of bacterial protein complexes initially requires the verification of protein complex formation *in vitro*. A summary of the experimental strategies to prepare and characterize protein complexes from bacteria is presented in Fig. 1

III. Preparation of Bacterial Protein Complexes

A. *General Expression and Affinity Purification Strategies*

The *in vitro* structure–function studies of any protein complex require its isolation in a form amenable to study. The isolation of protein complexes direct from bacterial cells through successive fractionation is a possible approach, with advantages including the isolation of relatively stable complexes amenable to electron microscopy (EM) or crystallization and the ability to identify new groups of interacting proteins. However, the use of recombinant methods offers significant advantages, especially for proteins of low abundance and for weakly associated protein complexes, including a higher level of protein expression and the potential to introduce fusion tags, either for detection or for affinity purification. The size and location of tags, however, must be kept in mind, in as much as they could negatively impact protein complex formation or stability.

With regard to the application of strategies for the expression of protein complexes, there are several specific issues that require careful consideration. When the individual protein components are well expressed and soluble, one can reconstitute the complex using individual protein components. There are different strategies that can be adopted, and it may be necessary to work through several of these in order to find the most appropriate one for the particular system of interest. The most straightforward is to mix together the purified proteins and isolate the complex by size exclusion chromatography (SEC). Another approach is to mix the cells isolated from individual bacterial cultures that were used to express the two (or more) proteins that form the complex. The cells can be combined, lysed by a variety of means, and the complex purified, for example, using an affinity tag. This approach can be especially powerful if the individual protein components are expressed with different affinity tags, allowing different, sequential affinity chromatographic steps to be employed in isolation of the complex. This method at the same time allows purification of the complex from any excess of one of the partners. Other variations on this approach can also be adopted, for example, combining individual bacterial lysates for the proteins being expressed or purification of one of the components by affinity chromatography, followed by incubation of the resin containing the purified protein with a lysate of the other protein (with or without a tag or with a different affinity tag) in order to capture one or more additional proteins.

In many situations, the above described approaches are not plausible. This is the case when one or more of the individually expressed protein components of a complex are insoluble or poorly soluble and unstable and have a tendency to precipitate. In this case, co-expression of the appropriate protein-binding partner can serve to stabilize and in some cases improve the solubility behavior of the protein of interest, providing the only way to obtain a viable complex (Tolia and Joshua-Tor, 2006). There are two approaches commonly used for co-expression: insertion of multiple targets in the same vector or co-transformation with compatible plasmids with different resistance genes and origins of replication, each carrying a single construct. Both approaches have certain advantages and disadvantages.

Several bi- or multicistronic vectors for co-expression of proteins have been described (Scheich *et al.*, 2007; Selleck and Tan, 2008). We have been using the bicistronic pET-Duet vector (Novagen) or a combination of up to four compatible versions of pET-Duet plasmids that permit simultaneous expression of up to eight different proteins (Kim *et al.*, 2006). One advantage of the co-expression strategy from a single vector is that the stoichiometry of a pair of interacting proteins should be easier to control, in as much as both transcripts and corresponding proteins are from a single plasmid, abrogating any effect of plasmid copy number. A disadvantage, however, is that for each new variant of the same protein pair (domains, tags), a new clone must be independently generated.

Using two or more compatible plasmids to express individual proteins of a complex provides a powerful advantage when several combinations of constructs with or without different tags have to be matched together in co-expression trials in order to screen for those that give the best-behaved complex. However, since the various plasmids differ in their copy numbers, and therefore protein expression levels, the desired stoichiometry may not be achieved and an excess of one of the partners is very likely. *A priori*, it is difficult to predict whether co-expression is the best strategy, although if the sequence of interest is predicted to have significant unfolded regions, this would serve as an important indicator favoring co-expression. The unfolded regions can be quite reliably predicted with programs such as DisProt (Sickmeier *et al.*, 2007), FoldIndex (Prilusky *et al.*, 2005), or GlobPlot (Linding *et al.*, 2003).

For a bacterial protein complex for which the genes are located adjacent to each other within an operon (polycistronic mRNA unit), another strategy is to clone the entire operon (or an appropriate segment) in one vector for co-expression. A catalog of transcriptional units in *E. coli* has been organized within the RegulonDB database (http://regulondb.ccg.unam.mx/ (Gama-Castro *et al.*, 2008)) and EcoCyc (http://ecocyc.org/ (Keseler *et al.*, 2009)).

IV. Characterization of Protein Complexes

A. *Size Exclusion Chromatography*

SEC provides several advantages in the preparation of a protein complex. It is often a powerful purification step prior to crystallization screening, removing protein aggregates, and polishing the protein sample, while at the same time permitting buffer exchange to the final buffer used for crystallization. As a means of characterizing protein complexes, it offers a variety of information. Under conditions that the column is properly calibrated, it gives an apparent molecular mass and, therefore, an approximate stoichiometry for the complex under study. When used sequentially, that is, through re-injection of the main elution peak or analysis of the same sample at different points in time, it offers information on the stability of the protein complex. It also offers some level of dynamic information, in that high-affinity, strongly interacting protein complexes are expected to elute with sharper, better defined peaks than those complexes in which there are fast on–off kinetics among the interacting proteins. In the later case, the peaks from SEC experiments are often observed to be broad or contain distinct leading or lagging shoulders, indicative of multiple components in solution. It is necessary, however, to exercise some care in the analysis of data for protein complexes from SEC. One issue is the effect of total protein concentration on complex formation and apparent stoichiometry – it is often desirable to analyze a protein complex at both higher (15–20 mg/ml) and lower (2–5 mg/ml) protein concentrations, in order to detect differences in behavior. Weaker complexes may only be stable in SEC experiments when the starting protein concentrations are relatively high. The second consideration is the need to pay careful attention to the individual masses of the component proteins; proteins that form dimers or larger homo-oligomers may mask, or be misinterpreted, as protein complexes themselves. Careful analysis of elution fractions by native PAGE and SDS-PAGE can clarify this issue.

B. *Dynamic Light Scattering*

Dynamic light scattering (DLS) offers a complementary approach to SEC for the characterization of protein complexes. Like SEC, it can also be used to determine the apparent molecular masses of protein complexes and is a useful tool for establishing complex stability. Compared to SEC, DLS uses relatively little protein, depending on the specific instrument. Using the DynaPro Plate Reader (Wyatt Technology Corp., Santa Barbara, CA; http://www.lightscattering.com), a volume of 40 μl at a concentration of 1 mg/ml is normally sufficient and can be recovered following the experiment. Using such an instrument, it is possible to screen a variety of buffers, pH values, ionic strength conditions, and ligands or other additives in order to establish under what conditions the complex is stable or not. This information can in turn be used to design alternative lysis buffers for purification or protein buffers for crystallization screening. It is beneficial to determine the solution behavior of the protein complex by DLS at relatively high protein concentration, compatible with that to be used for crystallization, in order to evaluate its propensity for aggregation.

C. *Native PAGE*

Like SEC and DLS, native PAGE can be used to assess the behavior and stability of a protein complex. Native gels can be especially powerful, as they require even less protein than does DLS (as little as a few micrograms), and many conditions, for example, as many as 30, can be tested and evaluated simultaneously. As with DLS, various buffers and additives can be incubated with the protein complex in order to determine (a) if both proteins co-migrate as a complex and (b) to visualize formation of aggregates or other forms of protein heterogeneity that may negatively impact crystallization screening. Once the "best" conditions have been established, an optimized buffer for purification and crystallization screening can be formulated, in part, based on these results.

D. *Isothermal Titration Calorimetry*

Isothermal titration calorimetry (ITC) is a technique that allows one to quantify the affinity of protein–protein interactions. Unlike surface plasmon resonance (SPR), the measurements involve both interacting proteins in solution, thereby avoiding potential artifacts arising from surface immobilization and protein labeling. Also, the stoichiometry of the binding reaction is measured directly during the experiment. A specific limitation of ITC is that the useful range of accurately measured affinities is somewhat narrower, typically with K_d in the range 0.1–10 μM, as compared with SPR. As well, typical ITC experiments require larger quantities of protein, though this problem is somewhat alleviated with the recently released iTC200 instrument from GE/MicroCal (GE Healthcare, Piscataway, NJ; http://www.microcal.com/), which uses a smaller, 200 μl sample cell than previous ITC instruments.

An ITC experiment is performed at a constant temperature while titrating one of the proteins (the "titrand") in the sample cell with the other protein (the "titrant") loaded within a syringe. After each addition of a small aliquot of the titrant, the heat released or absorbed in the sample cell is measured with respect to a reference cell filled with buffer. As the number of injections increases, the quantity of free protein available progressively decreases until the available binding sites become saturated. The binding constant (K_a), molar binding stoichiometry (N), molar binding entropy (ΔS^0), and molar binding enthalpy (ΔH^0) are determined directly from fitting the data, permitting calculation of the Gibbs free energy of binding. By repeating a titration at different temperatures, it is possible to determine the heat capacity change (ΔC_p) associated with the binding reaction, $\Delta C_p = \mathrm{d}\Delta H/\mathrm{d}T$.

The parameters ΔG, ΔH, ΔS, and ΔC_p are global properties of the system being studied and reflect contributions from the protein–protein binding reaction, conformational changes of the component molecules during association, as well as changes in molecule/solvent interactions and in the state of protonation. The relative magnitude of the change in binding enthalpy, ΔH, primarily reflects the strength of the interactions of the ligand with the target proteins, including van der Waals, salt bridges, hydrogen bonds, and electrostatic interactions, whereas the magnitude of the change in entropy, ΔS, is associated with solvent reorganization and other entropic contributions to binding (Leavitt and Freire, 2001). Typically, hydrophobic interactions involve the favorable burial of non-polar groups from contact with water and are considered to be entropic in nature. The change in heat capacity, ΔC_p, is the parameter in the ITC experiment with the most straightforward structural interpretation. This quantity is directly proportional to the change in the estimated amount of polar and apolar solvent accessible surface area buried on formation of the complex and, to a lesser extent, from changes in molecular vibrations (Freire, 1993). A negative value for ΔC_p indicates an increase in hydrophobic interactions upon binding.

Ideally, the protein concentration in the sample cell should be at least 10-fold higher than the expected K_d. In practice, titration experiments are often performed with 30–60 μM protein solution in the sample cell and a 10- to 20-fold higher concentration of titrant in the syringe to ensure a final titrant:titrand ratio of (2–4):1 in the reaction cell. Practically, this means that the more soluble and better behaved of the two proteins will usually be used in the syringe. Both protein samples are prepared in or dialyzed against the same buffer to minimize artifacts due to any differences in buffer composition, that is, heats of dilution. Maintaining an identical pH for the titrand and titrant solutions is particularly important. When studying protein–protein interactions, the best way to achieve an identical buffer composition is by passing both proteins through a size exclusion column using the same buffer prior to ITC measurements. While there are no special requirements for the buffers used, the high protein concentrations in ITC experiments make it desirable to use higher capacity (50–100 mM) buffers. This is especially important when using lyophilized peptides that are chemically synthesized, as they usually contain traces of strong acid. When one or both proteins require the presence of a reducing agent, it is recommended to use tris(2-carboxyethyl)phosphine (TCEP), which is more stable than dithiothreitol (DTT). DTT should be avoided in ITC buffers, as it often results in erratic baselines.

When using fusion proteins for ITC measurements, one should be aware of the possible interference of tags with the interactions being studied, such as occlusion of a binding site and non-specific binding. Special attention should be paid when using tags, which are not monomeric in solution, including the glutathione-*S*-transferase (GST) tag, as this may affect the stoichiometry of the binding event.

E. *Surface Plasmon Resonance*

SPR is a technique that yields real-time data on protein–ligand interactions and which can be used to guide or validate protein complexes for further structural studies (Masson *et al.*, 2000; McDonnell, 2001; Piliarik *et al.*, 2009). Typically, one binding partner called the ligand is immobilized on the surface of a microfluidic cartridge coated with an inert organic polymer matrix as sold by GE/Biacore (http://www.biacore.com). Then, a solution of the binding partner (the analyte) is injected at a constant flow rate over the surface. The SPR response is proportional to the mass at the surface, a feature that is used to monitor interactions between molecules in real time. An SPR sensorgram typically consists of a baseline phase (buffer flowing before injection of the analyte), a binding phase (during the injection), and a dissociation phase (after injection), when again buffer flows over the surface. The binding phase depends on both the association and dissociation kinetics, whereas the dissociation phase depends only on the dissociation kinetics. By injecting different concentrations of the analyte, typically over three orders of magnitude around the K_d, the equilibrium, and kinetic rate constants can be determined. SPR chips typically have four independently monitored flow cells on which different ligands can be immobilized and over which a single analyte solution can be flowed. One flow cell should be used as a control to account for bulk refractive index changes and non-specific adsorption of the analyte on the surface.

Although less than 1 μg of protein ligand is normally needed for one immobilization, identifying conditions for the immobilization of the protein remains one of the most crucial steps when designing an SPR experiment. The protein should be immobilized in a way that does not impede its interaction with a potential binding partner. Several immobilization procedures have been proposed. Proteins can be covalently coupled to a surface using thiol or amine-based chemistry. The former often requires engineering of a solvent-exposed cysteine in the protein ligand, whereas the latter involves EDC/NHS-mediated coupling through the side-chain ϵ-amino group of lysines or amino-terminal groups. The protein should be dissolved in a non-reactive buffer (no thiols or amines such as DTT or Tris, respectively) at a pH and salt concentration that allow the protein to interact electrostatically with the surface. However, the amine-coupling procedure can inactivate a protein and can lead to heterogeneity in the mode of ligand immobilization if the ligand has many exposed lysines, which can in turn affect determination of both stoichiometry and affinity. Alternatively, affinity tags such as hexa/deca-histidine, biotin, or GST can be used to immobilize the ligand on Ni-NTA, streptavidin, or anti-GST antibody-coated SPR chips, respectively. Because anti-GST antibody surfaces can be easily regenerated and the immobilization of GST-fusion proteins does not require covalent modification of the ligand, it remains one of the most popular SPR immobilization methods for investigating protein–protein interactions. The downside of using GST is that this protein is a dimer, which can lead to artificially enhanced affinities due to avidity effects.

The main advantages of SPR for the characterization of protein complexes are the wide range of affinities that can be determined ($K_d < 10^{-3}$ M) and the small amount of protein required for conducting an experiment. The highest concentration of analyte required to accurately measure a K_d should be 10–20 times greater than the dissociation constant in order to reach saturation of the ligand during the binding phase. A single injection requires 20–300 μl depending on the flow rate (10–50 μl/min) and the time required for the binding phase to reach a steady state (30–300 s). High flow rates (> 30 μl/min) are recommended to avoid rebinding and mass transport effects. In principle, any chemically inert buffer can be used, although it is common to use a neutral HEPES-buffered saline solution supplemented with a mild non-ionic detergent such as Tween-20 to prevent non-specific hydrophobic interactions. The availability of multiple flow cells on a single chip makes SPR especially suitable for rapidly evaluating the binding activity of multiple single-site mutant proteins with a second partner protein. In a single experiment, in addition to the control lane with GST or biotin alone, one can immobilize a wild-type protein and two mutant proteins and test them simultaneously with a single analyte concentration series.

F. *Characterization of Protein Complexes by NMR Spectroscopy*

Solution NMR spectroscopy has proven to be one of the most powerful techniques for characterizing protein–protein interactions, owing to the development of sensitive NMR instrumentation (higher magnetic fields, cryoprobes, etc.), NMR pulse sequences for assignments, and methods for the isotopic labeling of proteins (Clarkson and Campbell, 2003; Takeuchi and Wagner, 2006). Although solution NMR can be used to determine *de novo* structures of small protein complexes (<30 kDa), its scope is much wider as it can be used to map binding interfaces, monitor conformational changes, or determine kinetic or equilibrium binding constants. The range of K_d's that is typically measurable by NMR spectroscopy is between 10 μM and 10 mM.

The simplest NMR application consists of recording a one-dimensional proton NMR spectrum of a protein in an aqueous buffer. A good dispersion of relatively narrow resonances in the aliphatic (–2 to 4 ppm) and aromatic/amide (6–11 ppm) regions indicates that the protein is folded and not aggregated. However, a good deal more information can be obtained through the use of isotopic enrichment. Typically, NMR-detectable isotopes such as ^{15}N and ^{13}C are incorporated into recombinant proteins by growing plasmid-bearing bacteria in a minimal medium supplemented with uniformly labeled $^{15}NH_4Cl$ and/or ^{13}C-D-glucose. The protein can then be purified as usual and concentrated to greater than 100 μM in a low-salt acidic buffer (pH < 7.0) to record multidimensional heteronuclear experiments such as the ^{1}H-^{15}N HSQC.

Kinetic and mechanistic information can be gained through the use of NMR titrations. When a ligand is added in a stepwise fashion to a labeled protein, the position and intensity of a resonance at different molar ratios indicate the exchange rate regime (fast or slow) between the bound and unbound state of the labeled protein. These NMR titrations can also be used to quantify equilibrium binding constants between interaction partners (if the system is in a fast exchange regime) and to determine the stoichiometry of a complex. NMR offers the unique potential for characterizing the affinities of separate binding sites on a protein in a single experiment without recourse to mutagenesis studies, with every resonance-acting as a "probe" of a local binding event. For larger complexes, additional protein deuteration is required. TROSY experiments for amide signal assignment in combination with the cross-saturation technique (Nakanishi *et al.*, 2002; Takahashi *et al.*, 2000) provide high-quality definition of protein–protein interfaces.

NMR is particularly useful for studies of transient and ultraweak complexes, which are usually hard to crystallize. In addition to the ^{1}H-^{15}N HSQC experiment, which allows detection of weak yet specific interactions, an NMR method, transferred nuclear Overhauser spectroscopy, is particularly suitable for studies of weak complexes as it relies on fast exchange between free and bound states. This approach is applicable to systems in which one partner is small (e.g., a peptide). NMR methodology as applied to study weak interactions has been described in recent reviews (Prudencio and Ubbink, 2004; Vaynberg and Qin, 2006).

G. *Amide Proton/Deuterium Exchange*

Mass spectrometry combined with proton/deuterium exchange allows mapping protein–protein interaction interfaces (for reviews, see Lanman and Prevelige, 2004; Mandell *et al.*, 2005) and is especially useful for systems which have solubility or aggregation problems and which are difficult to crystallize. Amide $^1H/^2H$ exchange is followed by proteolytic digestion and detection by MS. Both MALDI and electrospray ionization techniques have been used successfully for such experiments. Since fast exchanging amides at the protein–protein interface are most useful for mapping the interfaces, the method is applicable for strong complexes (K_d below 100 nM; (Mandell *et al.*, 1998)). Alternatively, NMR spectroscopy can be used as a read-out technique. The advantage of NMR involves the detection of individual amide signals, but it requires isotope labeling (at least ^{15}N). Simpler exchange experiments, in which H_2O is replaced by ^2H_2O in the buffer, are too slow to capture surface amides. However, recent promising approaches based on using the aprotic solvent DMSO allow detection of protected surface amides through analysis of unfolded samples (Dyson *et al.*, 2008; Hoshino *et al.*, 2002). Such experiments can be particularly suitable for large oligomeric complexes made up of many small subunits.

V. Experimental Determination of the Structure of Protein Complexes

Due to the efforts of many individual research laboratories and more recently, structural genomics programs worldwide, the number of structures of individual proteins has increased very rapidly. Often, one or more of the components of a complex have a homolog with a known structure. The availability of this information permits, in addition to "high-resolution" methods, such as crystallography or NMR spectroscopy, the utilization of "low-resolution" methods including small-angle X-ray scattering (SAXS), EM, and computational molecular docking.

A. *Crystallography of Protein Complexes*

As with their preparation and characterization, the crystallization of protein complexes also presents some special challenges. The influence of the variety of ionic strength, pH, and chemical environments experienced by the protein complex during crystallization screening can mean practically that weaker complexes (those having high micromolar affinity) have a relatively low probability of yielding crystals. Crystallization trials should include, in the case of binary complexes, various molar ratios of the two proteins forming the complex. Protein complexes that form *via* extensive interactions with one another, that have high affinity, and that are relatively stable in solution, would appear to have the best chance of crystallizing. One of the practical outcomes of crystallization is that it is not unusual to obtain crystals of only one member of a protein complex included in the initial screening, and so it is necessary to check any crystals of the putative complex by careful washing and SDS-PAGE in order to confirm the crystal's content. There are few, if any, practical methods so far to evaluate the formation or stability of a given protein complex under conditions of sparse matrix crystallization screening. Efforts have been made, however, to design specific crystallization screens for protein complexes based on the published crystallization conditions of protein complexes contained within the RCSB Protein Data Bank (Radaev *et al.*, 2006) and have led to the commercially available ProPlex crystallization screen (http://www.moleculardimensions.com/). We have found the Protein Complex Suite from Qiagen (http://www1.qiagen.com/) to be especially useful. In those cases where the complex is found to be recalcitrant to efforts to crystallize it, the use of mutant proteins for one or more members of the complex, either with the goal of enhancing stability, for example, by the mutation of specific Cys residues, or using surface entropy mutagenesis (Cooper *et al.*, 2007) should be tried. As with individual proteins, attempting crystallization of the corresponding complex from various bacterial orthologs is another potentially useful strategy.

Structure determination involves the same procedures as for individual proteins. When the structure of a homolog of one of the proteins in the complex is known, the molecular replacement method has a good chance to lead to a full structure determination without resorting to SeMet substitution and using anomalous dispersion to solve the structure. In such cases, even low-resolution data (3–3.5 Å) may provide sufficient insight into complex formation if the model(s) of the protein (s) from a higher resolution study already exist. In such a case, the crystal of a complex diffracting to a resolution that would be too low for *de novo* structure determination can still be valuable. This is an important factor to keep in mind when assessing crystal usability.

B. NMR Spectroscopy of Protein Complexes

Because the amide ^{1}H and ^{15}N chemical shifts are influenced by different factors in their local environment, a ^{1}H-^{15}N HSQC correlation spectrum effectively provides a fingerprint of the folded state of a protein. Backbone assignments of these spectra can be obtained for proteins up to 25 kDa (or even greater size if the protein is deuterated) by recording triple-resonance spectra that correlate sequential and intra-residue $^{1}H^{N}$, ^{15}N, and $^{13}C^{\alpha}/^{13}C^{\beta}$ signals that can be assigned to specific amino acids in the protein sequence (Muhandiram and Kay, 1994). Once assignments are obtained, the ^{1}H-^{15}N HSQC spectra of a ^{15}N-labeled protein, which shows one peak for every backbone amide group in the protein, can be used to monitor its interaction with any unlabeled ligand that is added to the labeled protein solution. An example of this method as applied to the interaction of a yeast protein with a peptide of its cognate binding partner is shown in Fig. 2. Chemical shift perturbations (peak displacement or broadening) arise from changes in the environment of the NMR nucleus and can be

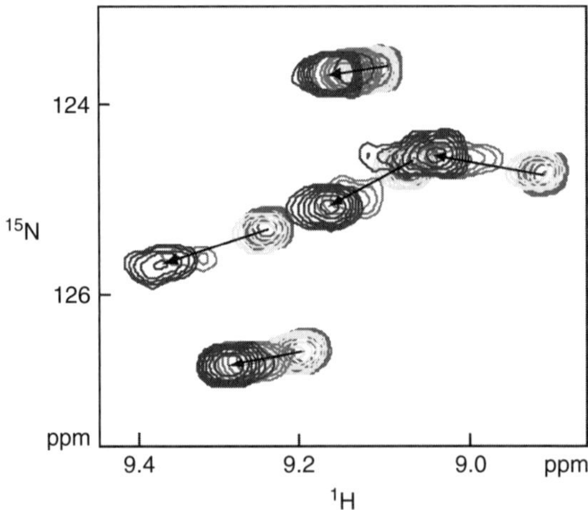

FIG. 2. NMR analysis of Ste50_RA/Opy2 peptide interactions. Overlay of regions of ^{1}H-^{15}N HSQC spectra for solutions containing a free RA domain (red) and increasing ratios of unlabeled Opy2 peptide titrated into ^{15}N-RA domain (orange, blue, and green, respectively). The arrows indicate displacement of selected ^{15}N-RA domain peaks. (See Color Insert.)

caused by direct protein–protein interactions or conformational changes induced by the binding event. These perturbations can then be mapped onto a protein structure to reveal interaction sites. Once the binding sites are known, the relative orientation of proteins in a complex can be readily determined using residual dipolar couplings (RDCs) measured in dilute liquid crystals (Bax *et al.*, 2001). RDCs can generally be fitted to structures determined in solution or to crystal structures, and can be used to derive more accurate structural models, as exemplified by the NMR-based docking of an acyl carrier protein to the acyl transferase enzyme LpxA (Jain *et al.*, 2004).

NMR spectroscopy can also provide dynamic information that can be used to further guide structural studies. Flexible termini or long loops can hinder crystallization and their removal can increase the likelihood of generating diffracting crystals (Page, 2008), but it is sometimes difficult to identify these regions with certainty through sequence analysis alone. The ^{15}N-^{1}H heteronuclear NOE experiment (Kay *et al.*, 1989) can be used to determine domain boundaries and flexible regions of proteins, as it depends on the rotational correlation time, which is itself a function of the folded state of the amide group. Fast-tumbling amide groups have negative heteronuclear NOE values whereas amide located in folded regions have NOE values above 0.6, a feature that can be used to delineate protein domains and to engineer recombinant protein constructs amenable to crystallization.

C. *Small-Angle X-ray Scattering*

In recent years there has been a revival in the use of SAXS to analyze protein shapes and conformations in solution (Koch *et al.*, 2003; Putnam *et al.*, 2007). The basic experimental design consists of exposing a protein solution to a collimated beam of hard X-rays and detecting X-rays scattered at a small angle. Data can be acquired at various synchrotron radiation facilities (SIBYLS beamline at the ALS, X33 at DESY, etc.) or using in-house instruments, such as the SAXSess Kratky camera (Anton Paar USA Inc, Ashland, VA; http://www.anton-paar.com/), the NanoStar U (Bruker AXS Inc., Madison, WI; http://www.bruker-axs.de/nanostar.html), and the PSAXS (Rigaku Corp., The Woodlands, TX; http://www.rigaku.com/index_en.html). The resulting rotationally averaged scattering curve is a scattering vector (q)-dependent intensity profile $I(q)$ that depends on the electron density contrast between the protein and the buffer, the shape and size of the protein, and intermolecular interferences (Glatter and Kratky, 1982). By recording SAXS profiles at different protein concentrations and by subtracting the buffer contribution with a blank measurement, it is possible to extrapolate the scattering curve corresponding to that of an infinitely diluted single particle. Any buffer can be used, but the salt concentration should be kept below 0.5 M to reduce background scattering. Glycerol may be added (up to 5%) to reduce radiation damage from synchrotron radiation. It is especially important that the buffer is the same in the blank and protein samples: ideally, the protein should be dialyzed and the dialyzate can be used as a blank. Gel filtration is a useful last step for SAXS analysis as it removes aggregates that can compromise data quality.

Various mathematical transformations can be applied to a scattering curve to evaluate the state of a protein complex in solution. The Guinier plot (q^2 vs. $\ln(I)$) should be linear at very low angles ($q^*R_g<1.3$) for a monodisperse solution of globular particles (Guinier and Fournet, 1955). Deviation from linearity in a Guinier plot indicates the presence of aggregates, which compromise the scattering curve, thus preventing further analysis (Fig. 3A). This quick and easy diagnostic can be run prior to embarking on crystallization trials with a concentrated protein solution, as the absence of aggregates correlates strongly with the probability of generating diffracting protein crystals (Jancarik *et al.*, 2004). If the Guinier plot is linear, the radius of gyration (R_g) and the forward scattering I_0 can be extracted from the slope and intercept of the plot, respectively (Fig. 3A). The latter can be used to calculate the molecular weight of the protein in solution as it depends solely on the mass and concentration of the scattering particle. This information can be used to determine the stoichiometry of a protein complex or oligomer. Finally, if the Guinier plot obtained at different protein concentrations shows a marked change in R_g and I_0/c values, it reveals the formation of concentration-dependent oligomers that can compromise SAXS analysis and crystallization experiments. Another useful diagnostic tool is the Kratky plot (q vs. I^*q^2), which is an indicator of the folded state of a protein (Putnam *et al.*, 2007). Folded globular proteins show a typical bell curve whose q value at the maximum can be used to determine an approximate molecular weight, whereas unfolded proteins show a plateau-shaped curve with no distinct peak.

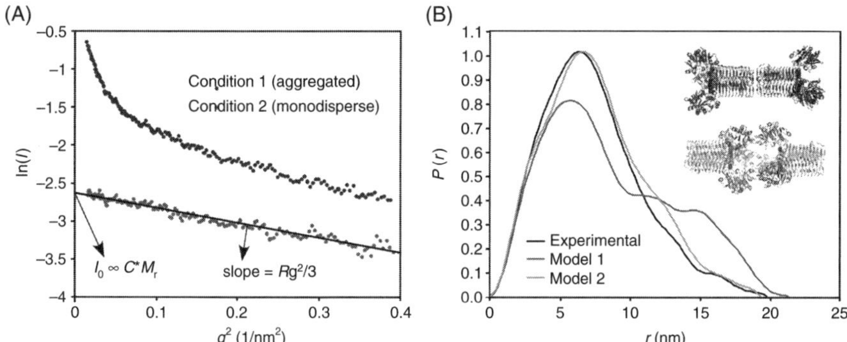

FIG. 3. Use of SAXS in structural genomics of protein complexes. (A) The Guinier plot, which is the square of scattering vector q ($4\pi \sin \theta/\lambda$) versus the natural logarithm of the intensity, can be used as a diagnostic of protein aggregation and size in solution. An example is shown here for a protein complex concentrated to 4 mg/ml in two different buffer conditions. A monodisperse globular particle solution should give a linear Guinier within the range $q_{\min}$ to $q_{\max}{*}R_g < 1.3$, as observed in condition 2. The slope enables calculation of R_g whereas the intercept enables the molecular weight to be determined if the concentration is known accurately. (B) The pair-density distribution function $P(r)$ allows the identification of a biological unit from a crystal structure that gives two possible hexameric forms (models 1 and 2). The experimental $P(r)$ function is much more similar to the hexamer #2, which implies that the protein forms hexamer of this type in solution. (See Color Insert.)

Although the information content of a scattering curve is relatively low, the combination of SAXS with other types of structural data can yield valuable information on the conformation of protein complexes in solution. The group of Dmitri Svergun at the EMBL-Hamburg has developed several data processing and analysis programs for SAXS (http://www.embl-hamburg.de/ExternalInfo/Research/Sax/software.html). A one-dimensional distance distribution function, $P(r)$, can be calculated from a scattering curve by inverse Fourier transformation using programs such as GNOM (Svergun *et al.*, 2001) or GIFT (Bergmann *et al.*, 2000). The $P(r)$ function provides a first glance at global structural features of a protein complex, that is, its maximum diameter ($D_{\max}$), the distance between globular domains, and whether it is spherical or elongated (Fig. 3B). Scattering curves and the $P(r)$ function can be computed directly from PDB coordinates and compared with experimental curves using the program CRYSOL (Svergun *et al.*, 1995). This procedure is useful to determine the oligomeric state of a protein complex in solution and to identify physiologically relevant binding interfaces in protein crystal structures (Fig. 3B). SAXS can also be used to ascertain conformational changes in a protein induced upon addition of small molecule inhibitors or cofactors. If the structures of individual components forming a complex are available, it is possible to perform a rigid-body docking of these structures to fit the experimental data. Finally, *ab initio* shape reconstruction can also be performed to model the shape of the scattering particle using an ensemble of dummy spheres with restraints that best represent the compactness of protein structures (Svergun *et al.*, 2001).

D. *Computational Molecular Docking with Experimental Restraints*

In a number of cases, the individual structures of interacting proteins are available, although efforts to co-crystallize the corresponding protein complex prove to be elusive. This is rather typical for transient interactions, where proteins do not possess high binding affinity. One of the possible ways to overcome this problem is the use of molecular docking to produce low-resolution complex models. The docking methods are classified into global methods based on geometric matching, Monte Carlo methods, and restraint-based High Ambiguity Driven biomolecular DOCKing (HADDOCK) approaches (Vajda and Kozakov, 2009). While the accuracy and reliability of predictions have improved significantly in recent years, the results of the first community-wide critical assessment of predicted interactions (CAPRI) experiment suggest that the addition of experimental information is critical to improve the accuracy of predicted models (Janin *et al.*, 2003).

The HADDOCK program makes ready use of experimental data to drive docking, unlike other approaches based on a combination of energetics and shape complementarity (Dominguez *et al.*, 2003). A variety of data, including NMR measurements, site-directed mutagenesis, and sequence conservation analysis, can be introduced as restraints for docking calculations. In particular, NMR is a valuable tool in obtaining HADDOCK restraints, including ambiguous interaction restraints (AIRs) from chemical shift mapping of interacting surfaces, unambiguous distance restraints in the form of intermolecular NOEs, and orientational restraints as RDCs. While NMR has limitations in terms of the size of proteins that can be easily studied, binding data can often be obtained for individual domains of multi-domain proteins.

The predicted structural model of the complex can be further confirmed using site-directed mutagenesis. Though the computational models lack the atomic resolution details of binding determinants, binding-deficient mutants can be designed and used by biologists to test the physiological importance of the protein complex in question, thereby providing important new information.

VI. Examples of Characterization of Bacterial Protein Complexes

In the following we provide several examples from our own work, illustrating various applications of the methodologies described in the preceding sections. These examples illustrate both the characterization of protein complexes and their structure determination by a variety of methods.

A. TtdA–TtdB: The L-tartrate Dehydratase of E. coli

E. coli utilizes L-tartrate as a fermentable carbon source under anaerobic conditions, *via* conversion to oxaloacetate. A hetero-tetrameric $\alpha_2\beta_2$ enzyme made up of two copies of TtdA and TtdB constitutes the *E. coli* L-tartrate dehydratase enzyme (Reaney *et al.*, 1993). The TtdA subunit contains a 4Fe–4S cluster, giving the protein a brown color and making the enzyme sensitive to oxygen. This enzyme is specific for the L-isomer of tartrate, with D-tartrate metabolized via the enzyme fumarase (Kim *et al.*, 2007a). Crystal structures are available for D-tartrate dehydratase (Yew *et al.*, 2006), but not for the L-specific enzyme. In order to obtain a structure for *E. coli* L-tartrate dehydratase, we have cloned, expressed, and purified the TtdA and TtdB subunits and attempted re-constitution of the enzyme by mixing and analysis by gel filtration chromatography. In this experiment, we did not succeed in obtaining the hetero-tetrameric complex. As an alternative strategy, we PCR-amplified the *ttdA-ttdb* genes in tandem and cloned them into several expression vectors. Co-expression of the two proteins together readily resulted in a well-behaved TtdA-TtdB complex. Efforts to crystallize this enzyme in the presence and absence of L-tartrate are ongoing. Our work on this complex illustrates the advantage, in this instance, of cloning and expression of both contiguous genes as one unit, as opposed to expressing, purifying, and mixing purified subunits individually.

B. *MnmG–MnmE: An Enzyme Complex Involved in tRNA Modification*

Bacteria extensively process and modify tRNA molecules, in part, to enhance codon–anticodon recognition, necessary for fidelity during translation (Bregeon *et al.*, 2001). One such modification involves the attachment of a 5-carboxymethylaminomethyl (cmnm) group onto the uridine base of U34 of selected tRNAs. A pathway responsible for synthesizing the cmnm group on U34, consisting of several enzymes, has been elucidated in *E. coli*. A FAD-dependent oxidoreductase, MnmG and GTPase MnmE function together to generate an intermediate form of the cmnm group subsequent to its methylation by MnmC. Using gel filtration chromatography, it has been shown that MnmG and MnmE form a 2:2 $\alpha_2\beta_2$ hetero-tetrameric complex *in vitro* (Yim *et al.*, 2006). Co-expression and purification of MnmG and MnmE in pET-Duet revealed that MnmE is able to co-purify with a His-tagged form of MnmG, although the quantity of MnmE appeared to be sub-stoichiometric, illustrating one of the limitations of co-expression in the case of weaker protein complexes. The MnmG interacting surface involves the C-terminal ~50–70 residues, as deletion of this region results in loss of complex formation with MnmE (Meyer *et al.*, 2008, M. Cygler, unpublished data). Additional studies, including alanine-scanning mutagenesis of the C-terminal region of MnmG, combined with protein–protein binding analysis by SPR or ITC, will be required to more precisely map the MnmG–MnmE interface.

C. *HypE–HypF and HypC–HypD: Protein Complexes Involved in [NiFe] Hydrogenase Maturation*

The *E. coli* genome encodes three hydrogenase systems in which the large hydrogenase subunit incorporates a [NiFe] metallocenter for its function (Forzi *et al.*, 2007). Hydrogenases make use of a number of maturation enzymes to synthesize the CO and CN ligands which are incorporated into the metal center. The two proteins required for synthesis of the cyano moiety are HypE and HypF (Jacobi *et al.*, 1992). HypF is an ATP-dependent enzyme, utilizing carbamoyl phosphate to generate a thiocarabamate moiety, which is transferred to the C-terminus of HypE (Reissmann *et al.*, 2003). This thiocarbamate then undergoes dehydration to a cyano moiety in an ATP-dependent reaction catalyzed by HypE (Blokesch *et al.*, 2004b). Previously, we have determined the crystal structure of *E. coli* HypE and characterized its interaction with HypF using several approaches (Rangarajan *et al.*, 2008). The affinity of the two proteins for one another was measured using both SPR and ITC, giving a K_d of approximately 400 nM. Gel filtration and scanning densitometry of SDS-PAGE gels were used to determine that the proteins form a 2:2 hetero-oligomeric complex. Depending on protein concentration, a 1:1 HypE–F complex also was observed. The SPR sensorgrams were consistent with a model in which a conformational change occurs in HypE–F upon complex formation. In order to define more precisely the binding site between HypF and HypE, a truncated version of HypF lacking residues 1–191 was PCR-amplified, expressed in *E. coli*, purified, and mixed with purified HypE protein. Analysis of the protein mixture revealed that this shorter version of HypF still forms a complex with HypE, and that the HypF deletion in the presence of HypE is apparently better behaved than the truncated version of the protein alone (Fig. 4A). The apparent mass of the complex by DLS, 100 kDa, is most consistent with a 1:1 stoichiometry of the two proteins (Fig. 4B). This apparent stoichiometry was further validated by SDS-PAGE analysis of fractions from SEC analysis of the complex (Fig. 4C). Efforts to further delineate the HypE–F interaction surface are ongoing.

In addition to complex formation between HypE and HypF, two other proteins, HypC and HypD, have been shown to form a complex together, as well possibly also with bound HypE (Blokesch *et al.*, 2004a). The function of the HypC–D complex is to presumably deliver Fe to the hydrogenase active site, as HypC binds Fe as a $[4Fe–3S]^{2+}$ cluster (Blokesch *et al.*, 2004a) and also forms a specific interaction with the precursor form of the hydrogenase III large subunit (Drapal and Bock, 1998). The molecular details of how HypC–D interact and donate the Fe atom to the large hydrogenase subunit remains unknown, although a model for the HypC–D complex has been proposed, based on the crystal structures of both proteins from *Thermococcus kodakaraen*sis (Watanabe *et al.*, 2007). We have focused on determining the co-crystal structure of *E. coli* HypC–

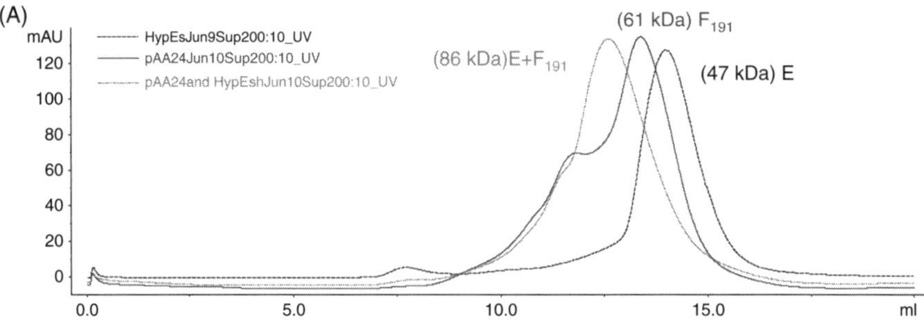

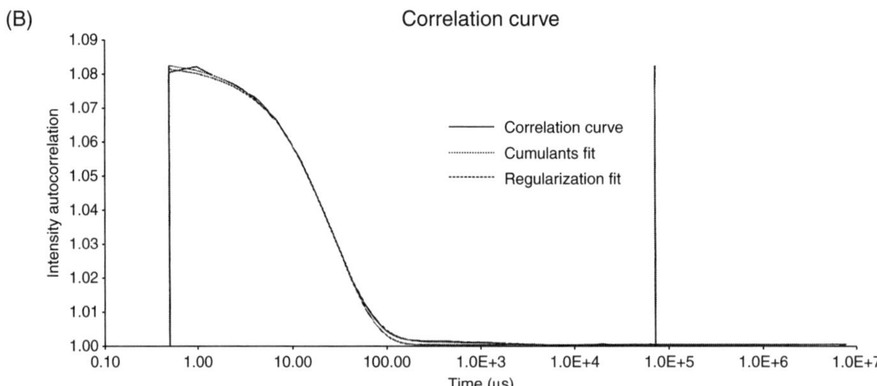

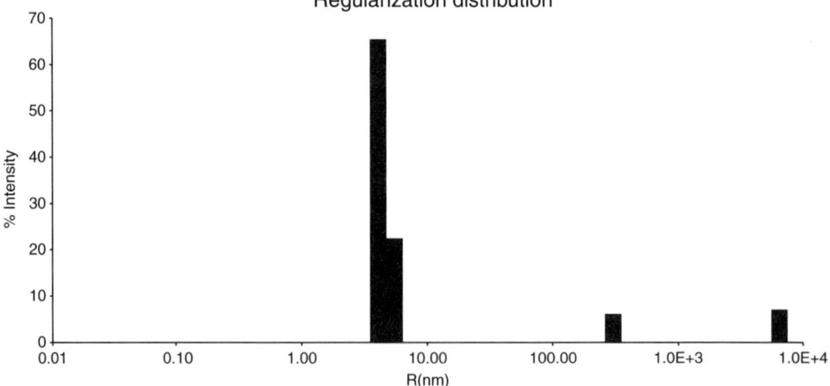

Fig. 4. (Continued)

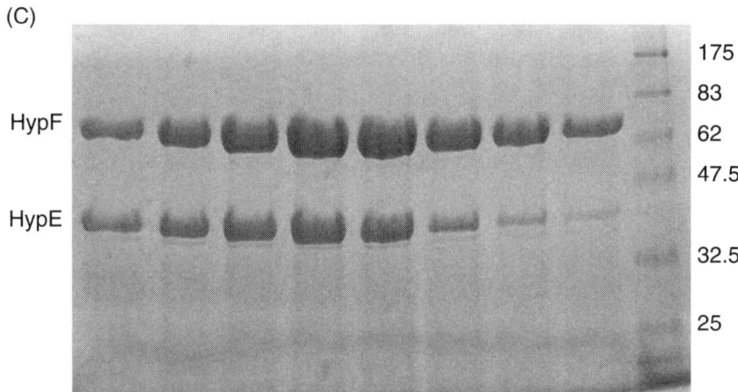

FIG. 4. Protein–protein complex between *E. coli* HypE and a deletion construct (Δ1–191) of *E. coli* HypF. (See Color Insert.)

D in order to address these questions. We have demonstrated complex formation between HypC–D by SEC, as well as by ITC. Titration of HypC into HypD yielded an association constant, K_a, of 200 nM, with an apparent 1:1 stoichiometry of the two proteins (Fig. 5). Based on the thermodynamic data, the interaction between the two proteins is primarily entropy driven, with the overall binding reaction being endothermic.

D. *YaeO–Rho: Inhibition of Rho-Dependent Transcription Termination*

Transcription termination is the process where a nascent RNA is released from its complex with RNA polymerase and the DNA template. In bacteria, two main mechanisms of transcription termination have been described. These mechanisms, commonly referred to as Rho-independent and Rho-dependent termination, are essential for the regulation of bacterial gene expression (Richardson and Greenblatt, 1996). Rho-dependent termination requires the presence of a hexameric helicase, Rho (Brown *et al.*, 1981), an essential transcription factor that binds nucleic acids at specific termination sites (rut), and translocates along the RNA until it reaches the transcription complex (Geiselmann *et al.*, 1993; Platt, 1994; Richardson, 1996). One of the Rho-specific inhibitors of transcription is the product of the *yaeO* gene, which reduces termination

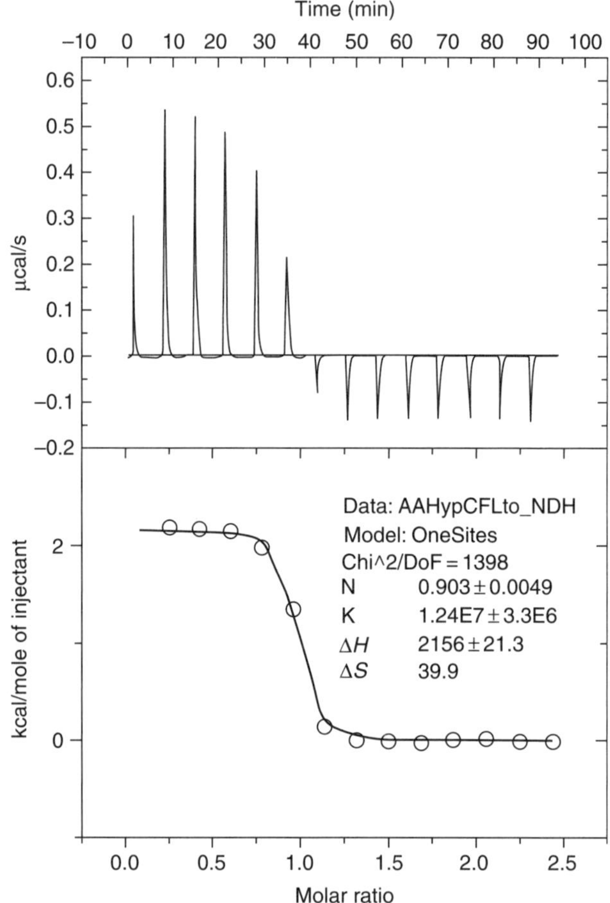

Fig. 5. ITC experiment to determine the association constant between the hydrogenase maturation proteins, HypC and HypD, from *E. coli*. The calculated association constant is 200 nM, with an apparent *n* value of 1.

in the Rho-dependent bacteriophage terminator tL1, and upstream the autogenously regulated gene *rho* (Pichoff *et al.*, 1998). Overexpression of YaeO can cause the pleiotropic suppression of conditional lethal mutations in cell division and heat shock genes, such as *ftsQ*, *ftsA*, *grpE*, *groEL*, and *groES* (Pichoff *et al.*, 1998).

We first determined the NMR solution structure of YaeO that revealed a topologically similar fold to that of the RNA-binding domain of small ribonucleoproteins (Sm-fold) (Gutierrez *et al.*, 2007). In order to understand the mechanism of transcription termination inhibition by YaeO, NMR experiments were used to characterize the interaction of YaeO with Rho *in vitro*. We used the N-terminal fragment (residues 1–130), referred as Rho130, that corresponds to the primary RNA binding site of Rho and has been shown to be a good model of Rho-oligonucleotide interactions (Briercheck *et al.*, 1998). The titration resulted in mapping the binding site of Rho130 on YaeO, which consists of the N- and C-termini, helix $\alpha 1$, and strands $\beta 3$, $\beta 4$, $\beta 5$, and $\beta 7$. These regions localize to one edge of the β-sandwich with clustered acidic residues. As the structure of Rho130 has been also determined (Briercheck *et al.*, 1998), we mapped the YaeO binding site on Rho130 that partially overlaps with the RNA binding surface, suggesting a mechanism of transcription termination inhibition.

As NMR titration data for the YaeO–Rho interaction was obtained for both proteins, a docking model of the complex was calculated using HADDOCK (Dominguez *et al.*, 2003). AIRs were derived from the NMR titration data by selecting residues with both the biggest chemical shifts and solvent accessibility. The resulting model is compatible with the hexameric Rho structure and reflects the charge complementarities of the interacting protein surfaces (Fig. 6B). This is consistent with *in vitro* binding results that show that the YaeO–Rho interaction is salt dependent and can be disrupted at high ionic strengths (Pichoff *et al.*, 1998). Importantly, the structural model was used to design the D14K, E19K, and E52K YaeO mutants that prevent inhibition of Rho activity using an *in vivo* β-galactosidase assay (Gutierrez *et al.*, 2007).

E. *SufBCD: A Protein Complex Involved in Bacterial Fe–S Cluster Synthesis Under Stress Conditions*

Iron–sulfur clusters are important metal cofactors in enzymes involved in a wide range of biological processes including respiration and the regulation of gene expression (Johnson *et al.*, 2005). Under most conditions, *E. coli* uses a general ISC pathway for the assembly of Fe–S proteins, but under conditions of oxidative stress bacteria can employ an alternative, Suf pathway (Nachin *et al.*, 2001). This pathway is well conserved in microorganisms and may play a role in bacterial pathogenesis

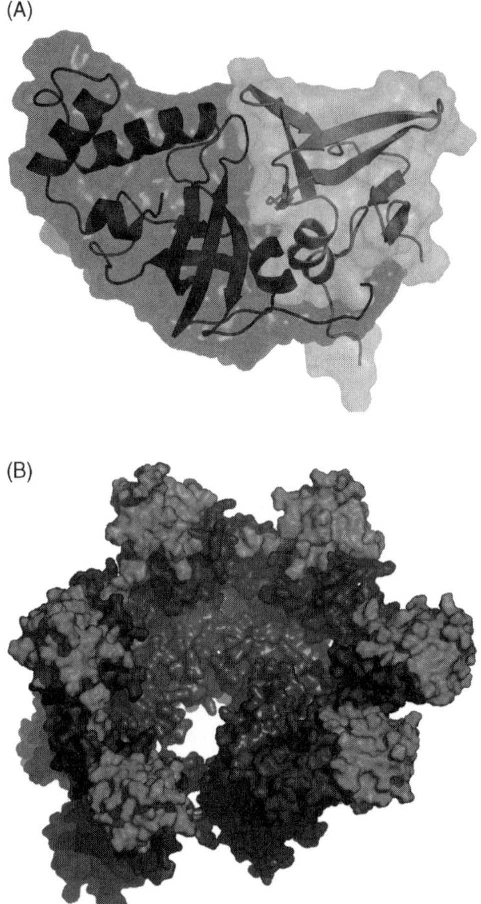

FIG. 6. (A) Model of the YaeO–Rho130 complex calculated with the program HADDOCK. YaeO is colored green and Rho130 is purple. (B) The YaeO/Rho130 model is compatible with the open-ring, hexameric form of Rho, accommodating six molecules of YaeO (in green) per Rho hexamer. The RNA-binding domain of Rho is colored purple and the ATP-hydrolysis domain is blue. (See Color Insert.)

by helping bacteria deal with the host immune response. The Suf system consists of a cysteine desulfurase (SufS) and five accessory proteins (Suf A, B, C, D, and E). SufBCD has been found to associate as a stable complex and to act synergistically with SufE to enhance the activity of the desulfurase SufS (Outten *et al.*, 2003). Our goal is to determine the three-dimensional structure of the SufBCD complex in order to examine the protein–protein interactions among the components of the cysteine def-sulfurase activator complex.

Initially the SufB, SufC, and SufD proteins were expressed separately in order to assemble the complex from purified components. While it proved possible to obtain pure samples of SufC and SufD using Ni-affinity chromatography, the final yield of SufC was low. We were able to improve the yield by combining the cell lysates from separately expressed SufC and SufD and co-purifying the complex. Unfortunately, expression of SufB on its own resulted in protein accumulating as insoluble aggregates, thwarting efforts to assemble the SufBCD complex from purified components. In an attempt to circumvent this problem, we decided to co-express all three components and purify the SufBCD complex. To accomplish this we cloned the entire *sufABCDSE* operon using an existing protocol (Outten *et al.*, 2003). Purification of the SufBCD complex was then carried out using a combination of anion exchange and SEC. Preliminary SDS-PAGE gels of these samples indicate the presence of proteins with molecular weights of 29, 47, and 55 kDa, values consistent with those of SufC, SufD, and SufB, respectively. By expressing the *suf* operon as a unit, we have been able to achieve the partial purification of the *E. coli* SufBCD complex. Additional purification steps will be required prior to crystallization trials.

F. *SdbA–CipA: The Mechanism of Cellulosome Cell-Surface Attachment*

The cellulosome is a large cell-surface bound multi-enzyme complex that synergistically degrades plant cell wall polysaccharides. First discovered in the thermophilic anaerobe *Clostridium thermocellum*, cellulosomes have since been identified in a variety of other anaerobic bacteria, ruminal bacteria, and anaerobic fungi (Bayer *et al.*, 2004). In general, the cellulosome can be divided into three modular protein components: (1) cell-surface proteins, (2) scaffold proteins, and (3) enzymes (Fig. 7A). Cellulosome assembly is mediated by conserved calcium-dependent protein–protein interaction modules called cohesins (Coh) and dockerins (Doc) (Bayer *et al.*, 2004). At least three types of Coh–Doc pairings exist, although they do not exhibit any cross specificity for other types (Ding

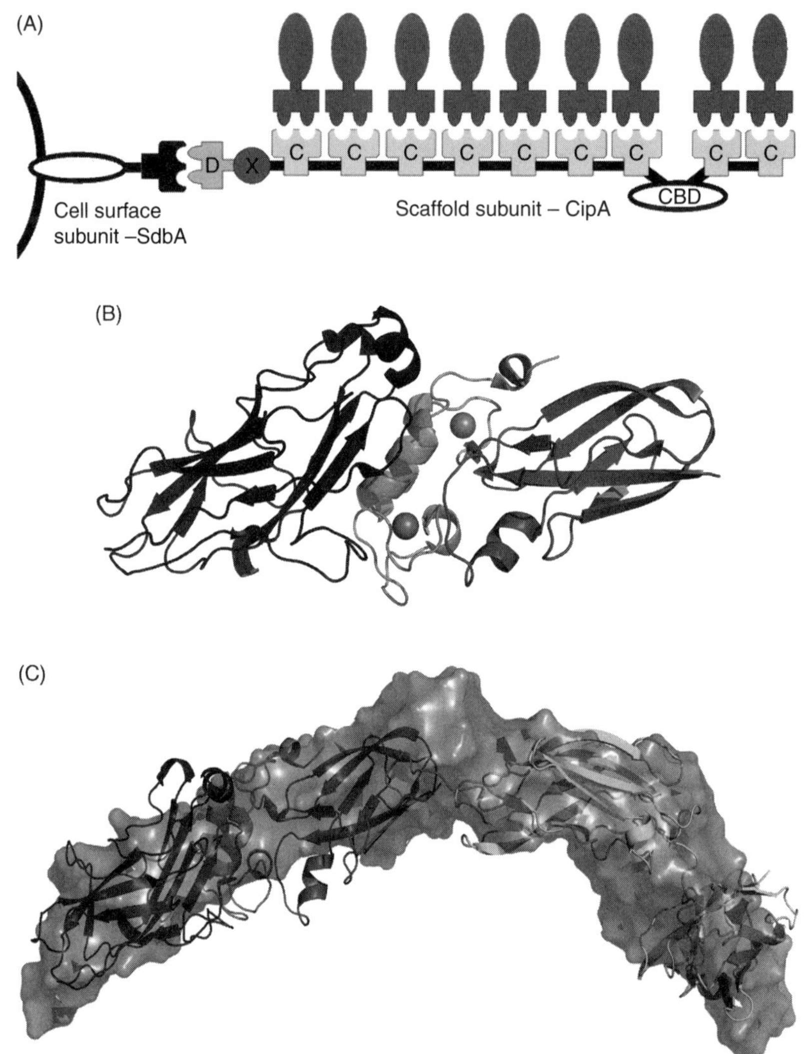

FIG. 7. (A) Cellulosome architecture: dark blue, type II Coh; green, type II Doc; pink, X-module; yellow, type I Coh; orange, type I Doc-containing enzymes. (B) Coh–DocX crystal structure: dark blue, type II Coh; green, type II Doc; pink, X-module; orange, calcium ions. (C) Crystal structures of Coh–DocX and two type I Cohs aligned with the SAXS structure of CohII–DocXCohICohI: gray, SAXS envelope; dark blue, type II Coh; green, type II Doc; pink, X-module; yellow, type I Coh; orange, calcium ions. (See Color Insert.)

et al., 2001). The scaffold subunit of *C. thermocellum*, CipA contains a C-terminal type II Doc module that anchors it to type II Coh-containing cell-surface proteins, nine type I Coh modules that bind to type I Doc-containing hydrolases, an X-module of unknown function, and a cellulose-binding domain (CBD) (Fujino *et al.*, 1993) (Fig. 7A). To date, three type II Coh-containing cell-surface proteins and more than 70 type I Doc-containing cellulases and hemicellulases have been identified in *C. thermocellum*, which offers a large degree of subunit plasticity (Bayer *et al.*, 2004).

We have characterized the mechanism of cellulosome cell-surface attachment in *C. thermocellum*, both biochemically and structurally, using several methods including ITC, DSC, SAXS, and X-ray crystallography. The type II Coh module from SdbA and a C-terminal fragment of CipA containing both the type II Doc and the X-module (DocX) were cloned, expressed, purified from inclusion bodies under denaturing conditions, and refolded (Adams *et al.*, 2004). ITC and DSC data indicate 1:1 binding and an ultra-high-affinity association constant (K_a) of 1.44×10^{10} M^{-1} (Adams *et al.*, 2006). We solved the crystal structure of this complex to 2.1 Å resolution using X-ray crystallography (Adams *et al.*, 2006) (Fig. 7B). The complex is elongated, with a highly hydrophobic interface and extensive hydrogen bonding between the X-module and both Doc and Coh modules (Adams *et al.*, 2006). The elongated structure allows the cellulosome to extend away from the surface of the cell to contact its extracellular substrate. Based on the observed X-module interactions, we propose a role for the X-module in Doc structure stability as well as in enhanced Coh recognition. The structure also provides a rationale for type I and type II Coh–Doc specificity, based on differences in Doc orientations, interface physicochemical properties, and X-module involvement. SAXS was used to gain further insight into the arrangement of neighboring modules. We solved the structure of the type II Coh of SdbA in complex with a fragment of CipA containing the type II Doc, the X-module, and two of the neighboring type I Cohs using SAXS (Fig. 7C). The SAXS envelope reveals an elongated, bent structure with no interactions evident between either of the type I Cohs and the other modules. This work illustrates a unique mode of cell-surface attachment, delineates a putative function for the previously uncharacterized X-module, and provides a rationale for the specificity of type I and type II Coh–Doc modules.

VII. Future Perspectives

The expression, purification, and crystallization of protein complexes pose challenging problems, both technically and scientifically. Tremendous effort will be required to validate high-throughput proteomics data and expand the list of tractable bacterial protein complexes that will be amenable to structural analysis. We can be certain that a great many protein complexes remain to be discovered. New methodologies, such as the use of whole-genome protein arrays (Oishi *et al.*, 2006), and *in vivo* split-protein complementation assays (Remy *et al.*, 2007) will contribute to discovering new protein–protein interactions. One approach that is likely to identify several new complexes is the systematic expression of bacterial operons, incorporating affinity tags at different points within the operon in order to "fish-out" interacting proteins. There is good reason to believe that functional linkages among proteins, including genetic linkage, can be used as a means to predict proteins that will interact (Kim *et al.*, 2007b). Knowledge of both predicted and experimentally observed disorder of individual proteins may offer an alternative avenue to identify new interacting proteins, as the presence of disorder correlates with some extent to a propensity to form a complex (Hegyi *et al.*, 2007). Despite the hurdles, many new and ultimately important biological insights will emerge from the structural analysis of protein complexes in the years ahead.

Acknowledgments

The authors wish to thank many of their colleagues whose work is described in this chapter, especially Eunice Ajamian Abdalin Asinas, Wayne Miller, Christine Munger, Ariane Proteau, Anna Rog, Rong Shi, as well many others for their contributions to refining the various methods described herein. This research was supported by a grant (GSP48370) from the Canadian Institutes of Health Research and by the National Research Council Canada. A.M.B. is a Canada Research Chair in structural biology. This is National Research Council of Canada publication no. 50671.

References

Adams, J. J., Jang, C. J., Spencer, H. L., Elliott, M., and Smith, S. P. (2004). Expression, purification and structural characterization of the scaffoldin hydrophilic X-module from the cellulosome of *Clostridium thermocellum*. *Protein Expr. Purif.* **38**, 258–263.

Adams, J. J., Pal, G., Jia, Z., and Smith, S. P. (2006). Mechanism of bacterial cell-surface attachment revealed by the structure of cellulosomal type II cohesin-dockerin complex. *Proc. Natl. Acad. Sci. USA.* **103**, 305–310.

Arifuzzaman, M., Maeda, M., Itoh, A., Nishikata, K., Takita, C., Saito, R., Ara, T., Nakahigashi, K., Huang, H. C., Hirai, A., Tsuzuki, K., Nakamura, S., taf-Ul-Amin, M., Oshima, T., Baba, T., Yamamoto, N., Kawamura, T., Ioka-Nakamichi, T., Kitagawa, M., Tomita, M., Kanaya, S., Wada, C., and Mori, H. (2006). Large-scale identification of protein–protein interaction of *Escherichia coli* K-12. *Genome Res.* **16**, 686–691.

Bahadur, R. P., Chakrabarti, P., Rodier, F., and Janin, J. (2004). A dissection of specific and non-specific protein-protein interfaces. *J. Mol. Biol.* **336**, 943–955.

Bax, A., Kontaxis, G., and Tjandra, N. (2001). Dipolar couplings in macromolecular structure determination. *Methods Enzymol.* **339**, 127–174.

Bayer, E. A., Belaich, J. P., Shoham, Y., and Lamed, R. (2004). The cellulosomes: Multienzyme machines for degradation of plant cell wall polysaccharides. *Annu. Rev. Microbiol.* **58**, 521–54. 521–554.

Bergmann, A., Fritz, G., and Glatter, O. (2000). Solving the generalized indirect Fourier transformation (GIFT) by Boltzmann simplex simulated annealing (BSSA). *J. Appl. Crystal.* **33**, 1212–1216.

Beyer, A., Bandyopadhyay, S., and Ideker, T. (2007). Integrating physical and genetic maps: From genomes to interaction networks. *Nat. Rev. Genet.* **8**, 699–710.

Blokesch, M., Albracht, S. P., Matzanke, B. F., Drapal, N. M., Jacobi, A., and Bock, A. (2004a). The complex between hydrogenase-maturation proteins HypC and HypD is an intermediate in the supply of cyanide to the active site iron of [NiFe]-hydrogenases. *J. Mol. Biol.* **344**, 155–167.

Blokesch, M., Paschos, A., Bauer, A., Reissmann, S., Drapal, N., and Bock, A. (2004b). Analysis of the transcarbamoylation-dehydration reaction catalyzed by the hydrogenase maturation proteins HypF and HypE. *Eur. J. Biochem.* **271**, 3428–3436.

Bregeon, D., Colot, V., Radman, M., and Taddei, F. (2001). Translational misreading: A tRNA modification counteracts a +2 ribosomal frameshift. *Genes Dev.* **15**, 2295–2306.

Briercheck, D. M., Wood, T. C., Allison, T. J., Richardson, J. P., and Rule, G. S. (1998). The NMR structure of the RNA binding domain of *E. coli* rho factor suggests possible RNA-protein interactions. *Nat. Struct. Biol.* **5**, 393–399.

Brown, S., Brickman, E. R., and Beckwith, J. (1981). Blue ghosts: A new method for isolating amber mutants defective in essential genes of *Escherichia coli*. *J. Bacteriol.* **146**, 422–425.

Butland, G., Babu, M., Diaz-Mejia, J. J., Bohdana, F., Phanse, S., Gold, B., Yang, W., Li, J., Gagarinova, A. G., Pogoutse, O., Mori, H., Wanner, B. L., Lo, H., Wasniewski, J., Christopolous, C., Ali, M., Venn, P., Safavi-Naini, A., Sourour, N., Caron, S., Choi, J. Y., Laigle, L., Nazarians-Armavil, A., Deshpande, A., Joe, S., Datsenko, K. A., Yamamoto, N., Andrews, B. J., Boone, C., Ding, H., Sheikh, B., Moreno-Hagelseib, G., Greenblatt, J. F., and Emili, A. (2008). eSGA: *E. coli* synthetic genetic array analysis. *Nat. Methods* **5**, 789–795.

Butland, G., Peregrin-Alvarez, J. M., Li, J., Yang, W., Yang, X., Canadien, V., Starostine, A., Richards, D., Beattie, B., Krogan, N., Davey, M., Parkinson, J., Greenblatt, J., and Emili, A. (2005). Interaction network containing conserved and essential protein complexes in *Escherichia coli*. *Nature* **433**, 531–537.

Carpousis, A. J. (2007). The RNA degradosome of *Escherichia coli*: An mRNA-degrading machine assembled on RNase E. *Ann. Rev. Microbiol.* **61**, 71–87.

Clarkson, J., and Campbell, I. D. (2003). Studies of protein-ligand interactions by NMR. *Biochem. Soc. Trans.* **31**, 1006–1009.

Collins, M. O., and Choudhary, J. S. (2008). Mapping multiprotein complexes by affinity purification and mass spectrometry. *Curr. Opin. Biotechnol.* **19**, 324–330.

Collins, R. F., Beis, K., Dong, C., Botting, C. H., McDonnell, C., Ford, R. C., Clarke, B. R., Whitfield, C., and Naismith, J. H. (2007). The 3D structure of a periplasm-spanning platform required for assembly of group 1 capsular polysaccharides in *Escherichia coli*. *Proc. Natl. Acad. Sci. USA* **104**, 2390–2395.

Cooper, D. R., Boczek, T., Grelewska, K., Pinkowska, M., Sikorska, M., Zawadzki, M., and Derewenda, Z. (2007). Protein crystallization by surface entropy reduction: Optimization of the SER strategy. *Acta Crystallogr. D* **63**, 636–645.

Datsenko, K. A., and Wanner, B. L. (2000). One-step inactivation of chromosomal genes in *Escherichia coli* K-12 using PCR products. *Proc. Natl. Acad. Sci. USA* **97**, 6640–6645.

Ding, S. Y., Rincon, M. T., Lamed, R., Martin, J. C., McCrae, S. I., Aurilia, V., Shoham, Y., Bayer, E. A., and Flint, H. J. (2001). Cellulosomal scaffoldin-like proteins from *Ruminococcus flavefaciens*. *J. Bacteriol.* **183**, 1945–1953.

Dominguez, C., Boelens, R., and Bonvin, A. M. (2003). HADDOCK: A protein-protein docking approach based on biochemical or biophysical information. *J. Am. Chem. Soc.* **125**, 1731–1737.

Drapal, N., and Bock, A. (1998). Interaction of the hydrogenase accessory protein HypC with HycE, the large subunit of *Escherichia coli* hydrogenase 3 during enzyme maturation. *Biochemistry* **37**, 2941–2948.

Driessen, A. J.M., and Nouwen, N. (2008). Protein translocation across the bacterial cytoplasmic membrane. *Ann. Rev. Biochem.* **77**, 643–667.

Dyson, H. J., Kostic, M., Liu, J., and Martinez-Yamout, M. A. (2008). Hydrogen–deuterium exchange strategy for delineation of contact sites in protein complexes. *FEBS Lett.* **582**, 1495–1500.

Enright, A. J., Iliopoulos, I., Kyrpides, N. C., and Ouzounis, C. A. (1999). Protein interaction maps for complete genomes based on gene fusion events. *Nature* **402**, 86–90.

Forzi, L., Hellwig, P., Thauer, R. K., and Sawers, R. G. (2007). The CO and CN(–) ligands to the active site Fe in [NiFe]-hydrogenase of *Escherichia coli* have different metabolic origins. *FEBS Lett.* **581**, 3317–3321.

Freire, E. (1993). Structural thermodynamics: Prediction of protein stability and protein binding affinities. *Arch. Biochem. Biophys.* **303**, 181–184.

Fujino, T., Beguin, P., and Aubert, J. P. (1993). Organization of a *Clostridium thermocellum* gene cluster encoding the cellulosomal scaffolding protein CipA and a protein possibly involved in attachment of the cellulosome to the cell surface. *J. Bacteriol.* **175**, 1891–1899.

Gama-Castro, S., Jimenez-Jacinto, V., Peralta-Gil, M., Santos-Zavaleta, A., Penaloza-Spinola, M. I., Contreras-Moreira, B., Segura-Salazar, J., Muniz-Rascado, L., Martinez-Flores, I., Salgado, H., Bonavides-Martinez, C., Abreu-Goodger, C.,

Rodriguez-Penagos, C., Miranda-Rios, J., Morett, E., Merino, E., Huerta, A. M., Trevino-Quintanilla, L., and Collado-Vides, J. (2008). RegulonDB (version 6.0): Gene regulation model of *Escherichia coli* K-12 beyond transcription, active (experimental) annotated promoters and Textpresso navigation. *Nucleic Acids Res.* **36**, D120–D124.

Gamba, P., Veening, J. W., Saunders, N. J., Hamoen, L. W., and Daniel, R. A. (2009). Two-step assembly dynamics of the *Bacillus subtilis* divisome. *J. Bacteriol.* **191**, 4186–4194.

Geiselmann, J., Wang, Y., Seifried, S. E., and von Hippel, P. H. (1993). A physical model for the translocation and helicase activities of *Escherichia coli* transcription termination protein Rho. *Proc. Natl. Acad. Sci. USA* **90**, 7754–7758.

Glatter, O., and Kratky, C. (1982). "Small Angle X-ray Scattering." Academic Press, London.

Goll, J., Rajagopala, S. V., Shiau, S. C., Wu, H., Lamb, B. T., and Uetz, P. (2008). MPIDB: The microbial protein interaction database. *Bioinformatics* **24**, 1743–1744.

Guinier, A., and Fournet, F. (1955). "Small Angle Scattering of X-rays." Wiley Interscience, New York.

Gutierrez, P., Kozlov, G., Gabrielli, L., Elias, D., Osborne, M. J., Gallouzi, I. E., and Gehring, K. (2007). Solution structure of YaeO, a Rho-specific inhibitor of transcription termination. *J. Biol. Chem.* **282**, 23348–23353.

Hegyi, H., Schad, E., and Tompa, P. (2007). Structural disorder promotes assembly of protein complexes. *BMC Struct. Biol.* **7**, 65.

Hoshino, M., Katou, H., Hagihara, Y., Hasegawa, K., Naiki, H., and Goto, Y. (2002). Mapping the core of the beta(2)-microglobulin amyloid fibril by H/D exchange. *Nat. Struct. Biol.* **9**, 332–336.

Jacobi, A., Rossmann, R., and Bock, A. (1992). The hyp operon gene products are required for the maturation of catalytically active hydrogenase isoenzymes in *Escherichia coli*. *Arch. Microbiol.* **158**, 444–451.

Jain, N. U., Wyckoff, T. J., Raetz, C. R., and Prestegard, J. H. (2004). Rapid analysis of large protein-protein complexes using NMR-derived orientational constraints: The 95 kDa complex of LpxA with acyl carrier protein. *J. Mol. Biol.* **343**, 1379–1389.

Jancarik, J., Pufan, R., Hong, C., Kim, S. H., and Kim, R. (2004). Optimum solubility (OS) screening: An efficient method to optimize buffer conditions for homogeneity and crystallization of proteins. *Acta Crystallogr. D* **60**, 1670–1673.

Janga, S. C., and Babu, M. M. (2009). Transcript stability in the protein interaction network of *Escherichia coli*. *Mol. Biosyst.* **5**, 154–162.

Janin, J., Henrick, K., Moult, J., Eyck, L. T., Sternberg, M. J., Vajda, S., Vakser, I., and Wodak, S. J. (2003). CAPRI: A critical assessment of predicted interactions. *Proteins* **52**, 2–9.

Jefferson, E. R., Walsh, T. P., Roberts, T. J., and Barton, G. J. (2007). SNAPPI-DB: A database and API of structures, interfaces and alignments for protein–protein interactions. *Nucleic Acids Res.* **35**, D580–D589.

Johnson, D. C., Dean, D. R., Smith, A. D., and Johnson, M. K. (2005). Structure, function, and formation of biological iron-sulfur clusters. *Annu. Rev. Biochem.* **74**, 247–281.

Kay, L. E., Torchia, D. A., and Bax, A. (1989). Backbone dynamics of proteins as studied by 15N inverse detected heteronuclear NMR spectroscopy: Application to staphylococcal nuclease. *Biochemistry* **28**, 8972–8979.

Keseler, I. M., Bonavides-Martinez, C., Collado-Vides, J., Gama-Castro, S., Gunsalus, R. P., Johnson, D. A., Krummenacker, M., Nolan, L. M., Paley, S., Paulsen, I. T., Peralta-Gil, M., Santos-Zavaleta, A., Shearer, A. G., and Karp, P. D. (2009). EcoCyc: A comprehensive view of *Escherichia coli* biology. *Nucleic Acids Res.* **37**, D464–D470.

Kim, O. B., Lux, S., and Unden, G. (2007a). Anaerobic growth of *Escherichia coli* on D-tartrate depends on the fumarate carrier DcuB and fumarase, rather than the L-tartrate carrier TtdT and L-tartrate dehydratase. *Arch. Microbiol.* **188**, 583–589.

Kim, S. M., Bowers, P. M., Pal, D., Strong, M., Terwilliger, T. C., Kaufmann, M., and Eisenberg, D. (2007b). Functional linkages can reveal protein complexes for structure determination. *Structure* **15**(9), 1079–1089. September 11, Ref Type: Abstract.

Kim, Y. G., Raunser, S., Munger, C., Wagner, J., Song, Y. L., Cygler, M., Walz, T., Oh, B. H., and Sacher, M. (2006). The architecture of the multisubunit TRAPP I complex suggests a model for vesicle tethering. *Cell* **127**, 817–830.

Kitagawa, M., Ara, T., Arifuzzaman, M., Ioka-Nakamichi, T., Inamoto, E., Toyonaga, H., and Mori, H. (2005). Complete set of ORF clones of *Escherichia coli* ASKA library (A Complete Set of *E. coli* K-12 ORF Archive): Unique Resources for Biological Research. *DNA Res.* **12**, 291–299.

Klem, T. J., and Davisson, V. J. (1993). Imidazole glycerol phosphate synthase: The glutamine amidotransferase in histidine biosynthesis. *Biochemistry* **32**, 5177–5186.

Kobe, B., Guncar, G., Buchholz, R., Huber, T., Maco, B., Cowieson, N., Martin, J. L., Marfori, M., and Forwood, J. K. (2008). Crystallography and protein–protein interactions: Biological interfaces and crystal contacts. *Biochem. Soc. Trans.* **36**, 1438–1441.

Koch, M. H., Vachette, P., and Svergun, D. I. (2003). Small-angle scattering: A view on the properties, structures and structural changes of biological macromolecules in solution. *Q. Rev. Biophys.* **36**, 147–227.

Kundrotas, P. J., and Alexov, E. (2007). PROTCOM: Searchable database of protein complexes enhanced with domain–domain structures. *Nucleic Acids Res.* **35**, D575–D579.

Lanman, J., and Prevelige, P. E. (2004). High-sensitivity mass spectrometry for imaging subunit interactions: Hydrogen/deuterium exchange. *Curr. Opin. Struct. Biol.* **14**, 181–188.

Lasserre, J. P., Beyne, E., Pyndiah, S., Lapaillerie, D., Claverol, S., and Bonneu, M. (2006). A complexomic study of *Escherichia coli* using two-dimensional blue native/SDS polyacrylamide gel electrophoresis. *Electrophoresis* **27**, 3306–3321.

Leavitt, S., and Freire, E. (2001). Direct measurement of protein binding energetics by isothermal titration calorimetry. *Curr. Opin. Struct. Biol.* **11**, 560–566.

Levy, E. D., Pereira-Leal, J. B., Chothia, C., and Teichmann, S. A. (2006). 3D complex: A structural classification of protein complexes. *PLoS. Comput. Biol.* **2**, e155.

Lin, C. C., Juan, H. F., Hsiang, J. T., Hwang, Y. C., Mori, H., and Huang, H. C. (2009). Essential core of protein–protein interaction network in *Escherichia coli*. *J. Proteome Res.* **8**, 1925–1931.

Linding, R., Russell, R. B., Neduva, V., and Gibson, T. J. (2003). GlobPlot: Exploring protein sequences for globularity and disorder. *Nucleic Acids Res.* **31**, 3701–3708.

Liu, G., Li, J., and Wong, L. (2008). Assessing and predicting protein interactions using both local and global network topological metrics. *Genome Inform.* **21**, 138–149.

Mandell, J. G., Baerga-Ortiz, A., Croy, C. H., Falick, A. M., and Komives, E. A. (2005). Application of amide proton exchange mass spectrometry for the study of protein-protein interactions. *Curr. Protoc. Protein Sci.* Chapter 20:Unit20.9.

Mandell, J. G., Falick, A. M., and Komives, E. A. (1998). Identification of protein-protein interfaces by decreased amide proton solvent accessibility. *Proc. Natl. Acad. Sci. USA* **95**, 14705–14710.

Masson, L., Mazza, A., and De, C. G. (2000). Determination of affinity and kinetic rate constants using surface plasmon resonance. *Methods Mol. Biol.* **145**, 189–201.

McDonnell, J. M. (2001). Surface plasmon resonance: Towards an understanding of the mechanisms of biological molecular recognition. *Curr. Opin. Chem. Biol.* **5**, 572–577.

Meyer, S., Scrima, A., Versées, W., and Wittinghofer, A. (2008). Crystal structures of the conserved tRNA-modifying enzyme GidA: Implications for its interaction with MnmE and substrate. *J. Mol. Biol.* **380**, 532–547.

Muhandiram, D. R., and Kay, L. E. (1994). Gradient-enhanced triple-resonance three-dimensional NMR experiments with improved sensitivity. *J. Magn. Reson. B* **103**, 203–216.

Nachin, L., El, H. M., Loiseau, L., Expert, D., and Barras, F. (2001). SoxR-dependent response to oxidative stress and virulence of *Erwinia chrysanthemi*: The key role of SufC, an orphan ABC ATPase. *Mol. Microbiol.* **39**, 960–972.

Nakanishi, T., Miyazawa, M., Sakakura, M., Terasawa, H., Takahashi, H., and Shimada, I. (2002). Determination of the interface of a large protein complex by transferred cross-saturation measurements. *J. Mol. Biol.* **318**, 245–249.

Oishi, Y., Yunomura, S., Kawahashi, Y., Doi, N., Takashima, H., Baba, T., Mori, H., and Yanagawa, H. (2006). *Escherichia coli* proteome chips for detecting protein–protein interactions. *Proteomics* **6**, 6433–6436.

Outten, F. W., Wood, M. J., Munoz, F. M., and Storz, G. (2003). The SufE protein and the SufBCD complex enhance SufS cysteine desulfurase activity as part of a sulfur transfer pathway for Fe-S cluster assembly in *Escherichia coli*. *J. Biol. Chem.* **278**, 45713–45719.

Page, R. (2008). Strategies for improving crystallization success rates. *Methods Mol. Biol.* **426**, 345–362.

Parrish, J. R., Yu, J., Liu, G., Hines, J. A., Chan, J. E., Mangiola, B. A., Zhang, H., Pacifico, S., Fotouhi, F., DiRita, V. J., Ideker, T., Andrews, P., and Finley, R. L., Jr. (2007). A proteome-wide protein interaction map for *Campylobacter jejuni*. *Genome Biol.* **8**, R130.

Patil, A., and Nakamura, H. (2005). Filtering high-throughput protein–protein interaction data using a combination of genomic features. *BMC Bioinformatics* **6**, 100.

Pichoff, S., Alibaud, L., Guedant, A., Castanie, M. P., and Bouche, J. P. (1998). An *Escherichia coli* gene (yaeO) suppresses temperature-sensitive mutations in essential genes by modulating Rho-dependent transcription termination. *Mol. Microbiol.* **29**, 859–869.

Piliarik, M., Vaisocherova, H., and Homola, J. (2009). Surface plasmon resonance biosensing. *Methods Mol. Biol.* **503**, 65–88.

Platt, T. (1994). Rho and RNA: Models for recognition and response. *Mol. Microbiol.* **11**, 983–990.

Prilusky, J., Felder, C. E., Zeev-Ben-Mordehai, T., Rydberg, E. H., Man, O., Beckmann, J. S., Silman, I., and Sussman, J. L. (2005). FoldIndex: A simple tool to predict whether a given protein sequence is intrinsically unfolded. *Bioinformatics* **21**, 3435–3438.

Prudencio, M., and Ubbink, M. (2004). Transient complexes of redox proteins: Structural and dynamic details from NMR studies. *J. Mol. Recognit.* **17**, 524–539.

Putnam, C. D., Hammel, M., Hura, G. L., and Tainer, J. A. (2007). X-ray solution scattering (SAXS) combined with crystallography and computation: Defining accurate macromolecular structures, conformations and assemblies in solution. *Q. Rev. Biophys.* **40**, 191–285.

Radaev, S., Li, S., and Sun, P. D. (2006). A survey of protein-protein complex crystallizations. *Acta Crystallogr. D* **62**, 605–612.

Rain, J. C., Selig, L., De, R. H., Battaglia, V., Reverdy, C., Simon, S., Lenzen, G., Petel, F., Wojcik, J., Schachter, V., Chemama, Y., Labigne, A., and Legrain, P. (2001). The protein–protein interaction map of *Helicobacter pylori*. *Nature* **409**, 211–215.

Rangarajan, E. S., Asinas, A., Proteau, A., Munger, C., Baardsnes, J., Iannuzzi, P., Matte, A., and Cygler, M. (2008). Structure of [NiFe] hydrogenase maturation protein HypE from *Escherichia coli* and its interaction with HypF. *J. Bacteriol.* **190**, 1447–1458.

Reaney, S. K., Begg, C., Bungard, S. J., and Guest, J. R. (1993). Identification of the L-tartrate dehydratase genes (ttdA and ttdB) of *Escherichia coli* and evolutionary relationship with the class I fumarase genes. *J. Gen. Microbiol.* **139**, 1523–1530.

Reissmann, S., Hochleitner, E., Wang, H., Paschos, A., Lottspeich, F., Glass, R. S., and Bock, A. (2003). Taming of a poison: Biosynthesis of the NiFe-hydrogenase cyanide ligands. *Science* **299**, 1067–1070.

Remy, I., Campbell-Valois, F. X., and Michnick, S. W. (2007). Detection of protein–protein interactions using a simple survival protein-fragment complementation assay based on the enzyme dihydrofolate reductase. *Nat. Protoc.* **2**, 2120–2125.

Richardson, J. P. (1996). Structural organization of transcription termination factor Rho. *J. Biol. Chem.* **271**, 1251–1254.

Richardson, J. P., and Greenblatt, J. (1996). "*Escherichia Coli* and *Salmonella Thyphimurium*: Cellular and Molecular Biology," pp. 822–848. Ed FC Neidhardt. Washington, DC: American Society for Microbiology Press.

Rutherford, S. T., Villers, C. L., Lee, J. H., Ross, W., and Gourse, R. L. (2009). Allosteric control of *Escherichia coli* rRNA promoter complexes by DksA. *Genes Dev.* **23**, 236–248.

Sato, S., Shimoda, Y., Muraki, A., Kohara, M., Nakamura, Y., and Tabata, S. (2007). A large-scale protein–protein interaction analysis in *Synechocystis* sp. PCC6803. *DNA Res.* **14**, 207–216.

Scheich, C., Kummel, D., Soumailakakis, D., Heinemann, U., and Bussow, K. (2007). Vectors for co-expression of an unrestricted number of proteins. *Nucleic Acids Res.* **35**, e43.

Selleck, W., and Tan, S. (2008). Recombinant protein complex expression in *E. coli*. *Curr. Protoc. Protein Sci.* Chapter 5:Unit 5.21.

Sickmeier, M., Hamilton, J. A., LeGall, T., Vacic, V., Cortese, M. S., Tantos, A., Szabo, B., Tompa, P., Chen, J., Uversky, V. N., Obradovic, Z., and Dunker, A. K. (2007). DisProt: The database of disordered proteins. *Nucleic Acids Res.* **35**, D786–D793.

Steitz, T. A. (2008). A structural understanding of the dynamic ribosome machine. *Nat. Rev. Mol. Cell. Biol.* **9**, 242–253.

Stenberg, F., Chovanec, P., Maslen, S. L., Robinson, C. V., Ilag, L. L., von, H. G., and Daley, D. O. (2005). Protein complexes of the *Escherichia coli* cell envelope. *J. Biol. Chem.* **280**, 34409–34419.

Su, C., Peregrin-Alvarez, J. M., Butland, G., Phanse, S., Fong, V., Emili, A., and Parkinson, J. (2008). Bacteriome.org an integrated protein interaction database for *E. coli*. *Nucleic Acids Res.* **36**, D632–D636.

Svergun, D., Barberato, C., and Koch, M. H.J. (1995). CRYSOL – A program to evaluate x-ray solution scattering of biological macromolecules from atomic coordinates. *J. Appl. Crystal.* **28**, 768–773.

Svergun, D. I., Petoukhov, M. V., and Koch, M. H.J. (2001). Determination of domain structure of proteins from x-ray solution scattering. *Biophys. J.* **80**(6), 2946–2953. June 1, Ref Type: Abstract

Takahashi, H., Nakanishi, T., Kami, K., Arata, Y., and Shimada, I. (2000). A novel NMR method for determining the interfaces of large protein-protein complexes. *Nat. Struct. Biol.* **7**, 220–223.

Takeuchi, K., and Wagner, G. (2006). NMR studies of protein interactions. *Curr. Opin. Struct. Biol.* **16**, 109–117.

Terradot, L., Durnell, N., Li, M., Li, M., Ory, J., Labigne, A., Legrain, P., Colland, F., and Waksman, G. (2004). Biochemical characterization of protein complexes from the *Helicobacter pylori* protein interaction map: Strategies for complex formation and evidence for novel interactions within type IV secretion systems. *Mol. Cell Proteomics* **3**, 809–819.

Tolia, N. H., and Joshua-Tor, L. (2006). Strategies for protein coexpression in *Escherichia coli*. *Nat. Methods* **3**, 55–64.

Tong, A. H., Lesage, G., Bader, G. D., Ding, H., Xu, H., Xin, X., Young, J., Berriz, G. F., Brost, R. L., Chang, M., Chen, Y., Cheng, X., et al., (2004). Global mapping of the yeast genetic interaction network. *Science* **303**, 808–813.

Vajda, S., and Kozakov, D. (2009). Convergence and combination of methods in protein-protein docking. *Curr. Opin. Struct. Biol.* **19**, 164–170.

Vaynberg, J., and Qin, J. (2006). Weak protein-protein interactions as probed by NMR spectroscopy. *Trends Biotechnol.* **24**, 22–27.

Watanabe, S., Matsumi, R., Arai, T., Atomi, H., Imanaka, T., and Miki, K. (2007). Crystal structures of [NiFe] hydrogenase maturation proteins HypC, HypD, and HypE: Insights into cyanation reaction by thiol redox signaling. *Mol. Cell* **27**, 29–40.

Wittig, I., and Schagger, H. (2008). Features and applications of blue-native and clear-native electrophoresis. *Proteomics* **8**, 3974–3990.

Wojcik, J., Boneca, I. G., and Legrain, P. (2002). Prediction, assessment and validation of protein interaction maps in bacteria. *J. Mol. Biol.* **323**, 763–770.

Yellaboina, S., Goyal, K., and Mande, S. C. (2007). Inferring genome-wide functional linkages in *E. coli* by combining improved genome context methods: Comparison with high-throughput experimental data. *Genome Res.* **17**, 527–535.

Yew, W. S., Fedorov, A. A., Fedorov, E. V., Wood, B. M., Almo, S. C., and Gerlt, J. A. (2006). Evolution of enzymatic activities in the enolase superfamily: D-tartrate dehydratase from *Bradyrhizobium japonicum*. *Biochemistry* **45**, 14598–14608.

Yim, L., Moukadiri, I., Björk, G. R., and Armengod, M. E. (2006). Further insights into the tRNA modification process controlled by proteins MnmE and GidA of *Escherichia coli*. *Nucleic Acids Res.* **34**, 5892–5905.

Yu, H., Chan, Y. L., and Wool, I. G. (2009). The identification of the determinants of the cyclic, sequential binding of elongation factors tu and g to the ribosome. *J. Mol. Biol.* **386**, 802–813.

STRATEGIES FOR THE CLONING AND EXPRESSION OF MEMBRANE PROTEINS

By CHRISTOPHER M. M. KOTH* AND JIAN PAYANDEH†

*Department of Structural Biology, Genentech, South San Francisco, California 94080
†Department of Pharmacology, University of Washington, Seattle, Washington 98195

Abstract

Despite the determination of thousands of high-resolution structures of soluble proteins, many features of integral membrane proteins render them difficult targets for the structural biologist. Among these, the most important challenge is in expressing sufficient quantities of active protein to support downstream purification and structure determination efforts. Over 190 unique membrane protein structures have now been solved, and noticeable trends in successful expression strategies are beginning to emerge. A number of groups have also explored high-throughput (HTP) methods for membrane protein expression, with varying degrees of success. Here we review the current state of expressing membrane proteins for functional and structural studies. We first survey successful methods

DOI: 10.1016/S1876-1623(09)76002-4

that have already yielded levels of membrane protein expression sufficient for structure determination. HTP methods are also examined since these aim to explore large numbers of targets and can predict reasonable starting points for many membrane proteins. Since HTP techniques may fail, particularly for certain classes of eukaryotic targets, detailed strategies for the expression of two prominent classes of eukaryotic protein families, G-protein-coupled receptors and ion channels, are also summarized.

I. Introduction

The first atomic structure of a membrane protein, the photosynthetic reaction center from *Rhodopseudomonas viridis*, was reported in 1985 (Deisenhofer *et al.*, 1985). Over the next decade, a small number of additional high-resolution structures of membrane proteins were solved and, as with the photosynthetic reaction center, most of these proteins were obtained from abundant natural sources (reviewed in Sakai and Tsukihara, 1998). In fact, almost every membrane protein for which an abundant natural source is known and a purification strategy has been established has proven amenable to structure determination. In 1992, the first crystal structures of outer membrane proteins expressed using recombinant DNA technology were reported (Cowan *et al.*, 1992). Six years later, the first recombinant α-helical membrane protein structures were reported (Chang *et al.*, 1998; Doyle *et al.*, 1998). This is in stark contrast to the situation for soluble proteins, where researchers have long exploited recombinant DNA technologies to overcome target protein expression (e.g., Blundell *et al.*, 1989). It is apparent that the expression of any membrane protein to sufficient levels in an active form represents a major bottleneck to high-resolution structural studies, and this can present significant technical and experimental challenges.

Many common and unique factors can limit the recombinant expression of membrane proteins. Problems common to the heterologous expression of any target include host codon bias, mRNA or protein stability, and cellular toxicity of the expressed protein (Sorensen and Mortensen, 2005). Many additional problems are unique to membrane proteins. First, the cellular machinery responsible for proper insertion of proteins into the membrane is limiting, and this may compromise overexpression efforts (Grisshammer, 2006; Wagner *et al.*, 2006, 2007). Second, predicting signal sequences or posttranslational modifications that may be required for proper membrane targeting, folding, and stability remains a difficult task (Li *et al.*, 2009). Third, the lipid composition of the expression host may not be compatible with the proper function, structure, or stability of the target protein. For example, yeast and prokaryotic membranes lack cholesterol while some human cell membranes contain approximately 30% cholesterol by molar ratio (Maxfield and Mondal, 2006). This is not a trivial concern since cholesterol has proven essential for the activity of many purified eukaryotic membrane proteins (Lagane *et al.*, 2000; Opekarova and Tanner, 2003). *Escherichia coli*, the workhorse of most soluble protein expression efforts, also lacks the enzymes required for the synthesis of phosphatidylcholine, yet this phospholipid is a major component of human cellular membranes (Opekarova and Tanner, 2003). For these and other reasons, it is not surprising that many eukaryotic membrane proteins require a eukaryotic host for the production of sufficient, active membrane protein to support structural studies.

Other factors may limit or direct the choice of expression host, particularly if structural studies are the experimental endpoint. If isotopic or heavy-atom derivative labeling is required for nuclear magnetic resonance (NMR) spectroscopy or crystallography studies, *E. coli* may be the system of choice, if not the only economically viable option. However, a given membrane protein target produced in different hosts or under varying expression conditions may yield a spectrum of samples of various quality and quantity (Auer *et al.*, 2001; Lundstrom *et al.*, 2006; Screpanti *et al.*, 2006; Tate *et al.*, 2003). Thus, although it is desirable to streamline cloning and expression protocols, maintaining flexibility is often necessary to find the most successful expression approach.

In addition to the selection of expression systems, one must also consider the choice and placement of affinity tags or fusion proteins. This is also not a trivial consideration, since affinity tags and their location can have dramatic effects on the expression level or stability of membrane proteins (Gordon *et al.*, 2008; Kawate and Gouaux, 2006; Lewinson *et al.*, 2008; Mohanty and Wiener, 2004). While strategies that have proven successful for other membrane proteins obviously provide reasonable starting points, this will remain an empirical endeavor for many targets.

We begin this review with a brief survey of the expression platforms and strategies that have already supported the successful structure determination of membrane proteins. These methods have provided considerable guidance for a number of high-throughput (HTP) efforts, which are also summarized. While many membrane proteins may be suitable for HTP expression and purification strategies, certain classes of targets seem to consistently present challenges, irrespective of their medical or pharmacological significance. G-protein-coupled receptors (GPCRs) and ion channels feature prominently among these, probably due to the structural and functional complexities of these dynamic molecules. We attempt to provide general guidance directed at overcoming expression obstacles for these two target classes.

II. From Past to Present: Practical Observations Regarding Membrane Protein Expression

To establish general starting points relevant for the production of most membrane proteins, a survey was conducted of the expression conditions that have been used to produce recombinant membrane targets for successful downstream X-ray crystallography or NMR spectroscopy (*data not shown*). Targets were identified by surveying the Membrane Proteins of Known 3D Structure Database (http://blanco.biomol.uci.edu/Membrane_Proteins_xtal.html), which comprises a list of all membrane protein structures determined to approximately 4.5 Å resolution, or better. Only structures of unique, non-monotopic α-helical membrane proteins produced by recombinant methods were examined. We focused on targets produced by recombinant methods since a very limited number of membrane proteins have an abundant natural source. For most membrane proteins, recombinant expression methods are necessary to obtain sufficient material to support structure determination efforts. Based on this survey and a review of the relevant literature, a number of guidelines for the cloning, tagging, and expression of membrane proteins can be recommended, as outlined in the following sections.

A. *Considerations for Cloning and Tagging*

The rate-limiting step in the production of any membrane target for structural studies is the generation of an expression clone that produces sufficient, active protein. This often requires the generation of many expression constructs and, possibly, the use of different expression hosts. Thus, cloning methods that permit the rapid and parallel production of expression clones should be considered (Marsischky and LaBaer, 2004). For example, the strategy employed in the Gateway® Cloning System (Hartley *et al.*, 2000; Walhout *et al.*, 2000) employs site-specific homologous recombination to produce an "entry vector" that is then used to shuttle the expression construct into a range of "destination vectors" (expression clones). The destination vectors are compatible with a wide array of expression host systems, affinity tags, and fusion proteins. For expression strategies requiring the construction of only a single destination vector, a simple one-step recombination strategy has also been developed (Fu *et al.*, 2008). The use of homologous recombination in combination with standardized positive-clone selection procedures is highly efficient and amenable to HTP clone generation. Expression vectors capable of directing protein production in different cell lines or multiple expression hosts can also be considered (Berrow *et al.*, 2007; Kost *et al.*, 2005), as can complete synthetic synthesis of expression clones. In short, advances in molecular biology have driven the process of expression clone generation to the point where it should no longer represent a bottleneck.

The choice and location of an affinity tag or fusion protein can have a profound effect on the expression and stability of any membrane protein target (Gordon *et al.*, 2008; Kawate and Gouaux, 2006; Mohanty and Wiener, 2004). For expression in *E. coli*, and most other hosts, the most frequently employed strategy is to place a stretch of at least six histidine residues (6xHis-tag) at the amino- or carboxyl-terminus of each target construct. When tackling a multi-subunit protein complex, each subunit of the complex should be prepared with or without a 6xHis-tag and then tested in combination with similarly prepared constructs of all other subunits (Locher *et al.*, 2002). If reasonable levels of protein expression can be achieved (i.e., ~0.5 mg/l of culture), purification using a general two-step procedure (i.e., capture of 6xHis-tagged proteins by immobilized metal affinity chromatography (IMAC) followed by size exclusion chromatography) typically results in samples of sufficient purity. Other affinity tags are viable options but are more rarely employed in *E. coli*, perhaps due to the increased costs associated with the affinity chromatography reagents (Waugh, 2005). Still, alternate tagging strategies can be considered in light of the crystallization of trace contaminants co-purifying during IMAC procedures (Psakis *et al.*, 2009; Stroud *et al.*, 2009; Veesler *et al.*, 2008).

A tagging strategy involving the incorporation of an antibody epitope may be recommended for the expression of membrane proteins in eukaryotic hosts. This is due, in part, to the considerable increase in contaminating host proteins that co-purify during a standard IMAC procedure. One possible strategy is the amino- and carboxyl-terminal tagging of constructs with different tags (i.e., amino-terminal FLAG epitope and carboxyl-terminal 6xHis-tag). This not only aids in purification but also improves the chances of initial target detection and purification in the face of possible amino- or carboxyl-terminal proteolytic processing (Li *et al.*, 2009). The addition of a so-called Rho-tag, comprising the first 20 amino acids of bovine rhodopsin (amino acid sequence: MNGTEGPNFYVPFSNKTGVV), has been found to dramatically boost the expression levels of several eukaryotic targets in insect and mammalian cells (Krautwurst *et al.*, 1998; Rice *et al.*, 2009). Therefore, the Rho-tag should be considered when approaching eukaryotic targets in a eukaryotic host. The expression of some membrane proteins can also be greatly enhanced if preceded by a leader sequence (Korepanova *et al.*, 2009). If targeting a homo- or hetero-oligomeric complex, differential tagging of subunits permits tandem-affinity purification and pull-down experiments which can help verify the structural integrity or stability of the complex under the experimental conditions. It is also important to note that the subsequent removal of an affinity tag (or fusion protein) using site-specific proteases may be inhibited in certain detergents or obscured in the protein construct under study (Lundback *et al.*, 2008; Mohanty *et al.*, 2003). Resolving these issues may require the alternate placement of affinity tags or the engineering of extended linkers.

Fusion proteins have been widely used for the production of soluble proteins but are more rarely cited in expression strategies for polytopic membrane proteins. This may stem from the increased size of fusion constructs, which can lead to improper membrane targeting or misfolding of the target protein. Nevertheless, some strategies have made use of fusion proteins and provide noteworthy examples. Following expression as carboxyl-terminal fusions to a leader sequence and maltose-binding protein in *E. coli*, two bacterial ligand-gated ion channels were crystallized and their structures solved (Bocquet *et al.*, 2009; Hilf and Dutzler, 2008, 2009). The amino-terminal-out topology of these channels likely contributed to their successful expression, and the strategy should be considered for other targets for which a similar topology of the amino-terminus is predicted. Indeed, using this approach, functional expression of diverse GPCRs in *E. coli* has been reported (McCusker *et al.*, 2007). Successful high-level expression using soluble cytosolic fusion proteins (e.g., glutathione-*S*-transferase, GST; thioredoxin, Trx) has also been demonstrated for some membrane proteins in *E. coli* (Ren *et al.*, 2009; Urban and Wolfe, 2005) and cell-free systems (Ishihara *et al.*, 2005). However, GST is known to form dimers and this can affect the structure or activity of the expressed membrane protein (Cacquevel *et al.*, 2008) or limit downstream assays. Although tenuous as a general membrane protein strategy, these types of fusion approaches have also permitted the isotopic labeling of some targets for NMR spectroscopy (Call *et al.*, 2006; Hu *et al.*, 2007; Oxenoid and Chou, 2005; Qin *et al.*, 2008; Schnell and Chou, 2008). The use of membrane-targeting or integrating fusion proteins can also be considered when attempting expression in *E. coli*, although there is limited evidence of functional reconstitution of eukaryotic targets using these approaches (Neophytou *et al.*, 2007; Roosild *et al.*, 2005).

The use of green fluorescent protein (GFP) as a fusion partner requires special mention since it can streamline key steps in the expression and detection of membrane protein targets. This is because whole-cell fluorescence measurements generally correlate with expression levels of target–GFP fusions (particularly for constructs in which GFP is fused to the carboxyl-terminus), detergent solubilization screening is greatly simplified, and proteins can be rapidly characterized using size exclusion chromatography without prior purification (Drew *et al.*, 2005, 2008; Geertsma *et al.*, 2008; Hammon *et al.*, 2009; Kawate and Gouaux, 2006; Newstead *et al.*, 2007). GFP can also be employed in selection strategies to identify host genes or target mutations that may permit increased expression levels or an increase in the stability of the target protein. Promising results have been achieved using a GPCR–GFP tagging strategy by screening against a panel of co-expressed chaperones (Link *et al.*, 2008). A similar approach was used to screen a transposon insertion library for gene disruptions in the expression host that resulted in higher expression of the target membrane protein (Skretas and Georgiou, 2008).

One further consideration is the codon usage or bias of the host versus the target gene. Some membrane protein structure determinations have absolutely required codon-optimized genes in order to express sufficient protein (e.g., MspA and PfAqp; Faller *et al.*, 2004; Newby *et al.*, 2008). *E. coli* strains expressing rare tRNAs are commercially available, and these have found widespread use (Brinkmann *et al.*, 1989; Del Tito *et al.*, 1995; Seidel *et al.*, 1992). However, a HTP screen in *E. coli* of 314 membrane proteins from diverse prokaryotic sources found no correlation between protein expression level and codon bias (Hammon *et al.*, 2009). Moreover, a recent genome-wide analysis has found that rare codons are often clustered between folding units (e.g., domains), and this may support the proper co-translational folding of proteins by slowing the translation rate (Zhang *et al.*, 2009). These (and other) coding features are frequently found in membrane protein genes, suggesting that important targeting and folding information may be present within native coding sequences (Prilusky and Bibi, 2009; Zhang *et al.*, 2009). Therefore, in addition to other considerations prior to synthetic gene construction (Kudla *et al.*, 2009), codon optimization should be considered if expression levels remain limiting. Ultimately, improved yields may require a series of "optimized" synthetic genes to be constructed and tested.

B. *Expression in Prokaryotic Hosts*

A short list of commonly utilized expression plasmids and expression systems is provided in Table I. However, for the vast majority of successful membrane protein structure determinations to date, the targets were of prokaryotic origin and *E. coli* was used as the expression host. Production in *E. coli* is fast and inexpensive, allowing many constructs to be screened quickly. *E. coli* cells are also among the most flexible for metabolic labeling strategies, and a wide range of expression vectors and cell strains are available (Jana and Deb, 2005; Makrides, 1996; Sorensen and Mortensen, 2005). The most productive HTP approaches for prokaryotic targets typically screen multiple homologs in different expression vectors using a variety of *E. coli* cell strains (Eshaghi *et al.*, 2005; Lewinson *et al.*, 2008; Surade *et al.*, 2006). The most common *E. coli* expression systems are those utilizing the T7 phage promoter (Studier *et al.*, 1990) with additional regulatory elements (i.e., as found in the popular pET vectors). Employing the so-called auto-induction method (Studier, 2005) can further circumvent toxicity that is often associated with leaky expression.

Table I
Commonly Employed Expression Plasmids Used for Structural Studies[a]

Host	Plasmid	Promoter	Tag(s)	Comments	Examples
Escherichia coli	pET-15b/19b	T7	Cleavable, N-terminal, His-6	Strong promoter, inducible	MscL (Chang *et al.*, 1998) CorA (Lunin *et al.*, 2006)
Escherichia coli	pET-28	T7	Cleavable, N- or C-terminal, His-6	Strong promoter, inducible	SNARE complex (Stein *et al.*, 2009) MgtE (Hattori *et al.*, 2007) PfAqp (Newby *et al.*, 2008)
Escherichia coli	pBAD	Arabinose	Cleavable, N- or C-terminal, His-6	Moderate promoter, inducible, tightly regulated	SecYEG (van den Berg *et al.*, 2004) Glt (Yernool *et al.*, 2004)
Escherichia coli	pQE	T5	Cleavable, N- or C-terminal, His-6	Moderate promoter, inducible, tightly regulated	MthK (Jiang *et al.*, 2002) KvAP (Jiang *et al.*, 2003) NaK (Shi *et al.*, 2006)

(continued)

Table I (*continued*)

Host	Plasmid	Promoter	Tag(s)	Comments	Examples
Yeast (*Pichia pastoris*)	pPICZ	*AOX1*	N- or C-terminal, c-myc/His-6	Strong promoter, inducible, tightly regulated; N-terminal α-factor secretion signal to help boost expression	Kv1.2 (Long *et al.*, 2005) Aqy1 (Fischer *et al.*, 2009) Aqp5 (Horsefield *et al.*, 2008) LTC_4 synthase (Martinez Molina *et al.*, 2007)
Yeast (*Shizosaccharomyces pombe*)	pESP	*nmt1*	C-terminal, His-6	Strong promoter, inducible	LTC_4 synthase (Ago *et al.*, 2007)
Yeast (*Saccharomyces cerevisiae*)	pUC18 (YEp351)	*PMA1*		Strong promoter	H^+-ATPase (Pedersen *et al.*, 2007)
Yeast (*Saccharomyces cerevisiae*)	pYES2.0	*GAL6*	Cleavable, N-terminal, His-10	Strong promoter	Aqp0 (Palanivelu *et al.*, 2006)
Yeast (*Saccharomyces cerevisiae*)	pYeDP1-8/10, pYeDP60	*GAL10*	Cleavable, N-terminal, BAD-domain	Strong promoter; BAD-domain aids affinity purification	Ca^{2+}-ATPase (Jidenko *et al.*, 2005)

Insect (Sf9, Hi5)	pBlueBac or pFastBac	Polyhedrin	None or N-terminal His-8	Strong, late viral promoter	ASIC1 (Jasti *et al.*, 2007) β_1-GPCR (Warne *et al.*, 2008) A_{2A}-GPCR (Jaakola *et al.*, 2008) Cx26 (Maeda *et al.*, 2009) $P2X_4$ (Kawate *et al.*, 2009)
Mammalian (COS cells)	p91023(B)	AdMLP		Strong, late viral promoter	Rhodopsin (Standfuss *et al.*, 2007)
Mammalian (HEK cells)	pEGFP-C1 (pNGFP-EU and pCGFP-EU)	CMV	N- or C-terminal His-8	Strong viral promoter	(Kawate and Gouaux, 2006)

Representative membrane protein targets are listed and should be referenced for additional construct design and expression details.

[a] In some cases the commercially available expression plasmids have been tailored for specific projects.

The formulation of the auto-induction media permits cell growth to very high densities and does not require monitoring of cultures prior to target protein induction. These factors tend to have significant impacts on membrane protein expression levels (Attrill *et al.*, 2009; Deacon *et al.*, 2008; Gordon *et al.*, 2008; Korepanova *et al.*, 2007). Further benefit may also be afforded by use of the C41 and C43 "Walker" expression strains (Miroux and Walker, 1996). These strains were selected based on their ability to overexpress membrane proteins and have recently been shown to act by slowing expression from the strong T7 promoter (Wagner *et al.*, 2008). The pBAD vectors are another popular membrane protein expression system and harbor a very tightly controlled arabinose promoter (Guzman *et al.*, 1995). While initial screening conditions can be standardized, titration experiments with the inducer arabinose should be performed for each pBAD construct. In some cases, expression optimization steps have been absolutely necessary to remove degradation products or other co-purifying contaminants prior to successful structure determination (e.g., GlpT and Na^+/H^+ antiporter; Auer *et al.*, 2001; Screpanti *et al.*, 2006). The above approaches have yielded the majority of prokaryotic membrane protein structures and the growing number of eukaryotic structures derived from *E. coli* produced material.

One other prokaryotic host merits consideration for membrane protein expression. A variety of prokaryotic and eukaryotic membrane proteins have been functionally expressed in *Lactococcus lactis*, with levels ranging from 0.1% to 30% of the total cellular membrane protein (Kunji *et al.*, 2003; Monne *et al.*, 2005). The cultivation of *L. lactis* is easy and inexpensive, and auxotrophic strains may allow for the labeling of proteins in defined media (Berntsson *et al.*, 2009). Well-established genetic methods and expression vectors are also available for *L. lactis*, where expression is generally driven by the tightly controlled *nisA* promoter. A medium-scale study by Surade *et al.* (2006) demonstrated that many of the prokaryotic targets that undergo rigorous expression efforts in *E. coli* can often be expressed in *L. lactis*.

C. *Expression in Yeast*

The yeasts *Pichia pastoris*, *Saccharomyces cerevisiae*, and *Schizosaccharomyces pombe* are champions of recombinant eukaryotic membrane protein production for high-resolution structure determination. Of the 26 crystal structures of eukaryotic proteins produced using recombinant methods, eight were expressed in *P. pastoris*, two used *S. cerevisiae*, and one was produced in *S. pombe*. Yeast expression systems are traditionally considered a compromise between the more expensive, time-consuming and labor-intensive insect or mammalian cell culture systems and the low-cost and flexibility of *E. coli*. Yeast hosts also offer the advantages of eukaryotic membrane-targeting machinery, chaperones, and the potential for faithful posttranslational modifications, in addition to the ease with which they can be cultured and genetically manipulated. It should be noted that yeast glycosylation patterns differ significantly from those of mammalian cells, although *P. pastoris* strains engineered to produce "humanized" glycosylation patterns have also been developed (Hamilton and Gerngross, 2007). In HTP settings, *S. cerevisiae* may provide a number of distinct advantages over *P. pastoris* including the availability of numerous expression strains and plasmids, as well as the ability to carry out functional complementation with mutant or deletion strains. Two HTP membrane protein expression screens in *S. cerevisiae* have reported promising results using the FGY217 strain or the W303-Δ*pep4* strain (both harbor deletions of the pep4p protease) and the galactose-inducible promoter from the *GAL1* gene (Li *et al.*, 2009; Newstead *et al.*, 2007). These report significant expression for ~70% of membrane proteins homologously expressed in *S. cerevisiae* (25/29 and 272/351, respectively) and a success rate of ~30% (4/14) for a limited number of exogenous targets from higher eukaryotes. One of these studies employed a carboxyl-terminal GFP-6xHis-tag fusion strategy, while the other used an amino-terminal FLAG and carboxyl-terminal 6xHis-tag tandem approach. In a smaller survey examining only seven membrane proteins (six from higher eukaryotes), *S. cerevisiae* and *P. pastoris* were found to be inferior to insect and mammalian cell systems (Eifler *et al.*, 2007). The lipid composition of yeast differs significantly from insect and mammalian cells, and this may adversely affect the functional expression of many target proteins. Nevertheless, robust functional expression of numerous GPCRs has been reported in *P. pastoris* strain SMD1163 using the inducible *AOX1* promoter and construct fusion to the α-factor signal sequence (Lundstrom *et al.*, 2006). Standard expression conditions in yeast are often found to be quite productive, and multiple formats have been introduced for HTP optimization and sample characterization (Andre *et al.*, 2006; Ito *et al.*, 2008; Newstead *et al.*, 2007; Sugawara *et al.*, 2009; Wedekind *et al.*, 2006; Zeder-Lutz *et al.*, 2006).

D. *Expression in Insect Cells*

Insect cell systems have been widely used for the production of soluble proteins for structural studies and technological advances have simplified the production and titration of viral stocks. Eleven recombinant eukaryotic membrane protein structures have now been determined using insect cells as the expression host. These include six ion channels and five GPCR structures. The membrane composition and protein processing machinery of insect cells are closer to that of mammalian cells than are yeast. In general, target genes are cloned with either an amino- or carboxyl-terminal affinity tag and downstream of the strong polyhedrin or p10 promoter, although a weaker promoter system may be advantageous for some targets (Higgins *et al.*, 2003). In some cases, addition of the aforementioned Rho-tag from rhodopsin, or a leader sequence, has been required for sufficient protein expression and downstream purification (Korepanova *et al.*, 2009; Rice *et al.*, 2009). Increases in functional expression levels may further be obtained by adding various DNA elements onto the construct, increasing the gene dosage, decreasing the growth temperature, co-expressing chaperones, or by using different cell strains (Akermoun *et al.*, 2005; Brock *et al.*, 2001; Higgins *et al.*, 2003; Kost *et al.*, 2005). Also, some engineered viruses are less susceptible to lysis upon prolonged infection times, and this may increase the overall yield of functional protein (Ho *et al.*, 2004). Incorporation of selenomethionine into protein produced in insect cells is possible and has already assisted in *de novo* membrane protein determinations (Jasti *et al.*, 2007; Kawate *et al.*, 2009; Maeda *et al.*, 2009). Of particular note is the relatively new "BacMam" technology which allows for the production of viral particles containing expression cassettes compatible with many mammalian hosts (Kost *et al.*, 2005). Thus, viral stocks produced in this system can also be used to test expression in mammalian cells.

E. *Expression in Mammalian Cells*

To date, recombinant expression in mammalian cells has been used for the high-resolution crystal structure determination of only one membrane protein, bovine rhodopsin, expressed in COS-1 cells (Standfuss *et al.*, 2007). Still, mammalian hosts may provide the highest levels of expressed *active* protein for many membrane targets (Eifler *et al.*, 2007;

Lundstrom *et al.*, 2006; Tate *et al.*, 2003), but there are only a few reports of high-level expression that could support structural studies or HTP endeavors (Hassaine *et al.*, 2006; Lundstrom *et al.*, 2006; Tate *et al.*, 2003). Medium- to large-scale transient transfection is perhaps the simplest means of testing the expression of membrane targets in mammalian hosts (Junge *et al.*, 2008; Kawate and Gouaux, 2006; Mancia and Hendrickson, 2007). Optimization of culture or media components is relatively simple and can often increase transfection efficiency and protein expression levels (Blasey *et al.*, 2000; Hassaine *et al.*, 2006; Kost *et al.*, 2005). Cell culturing technologies such as roller bottles, orbital shakers, and wave bags have also made these systems more accessible and simplify scale-up (Haldankar *et al.*, 2006; Singh, 1999). A wide variety of inducible promoters are available, which may address potential toxicity issues and/or maximize the levels of functional protein (Midgett and Madden, 2007). As already mentioned, the BacMam system can also permit the rapid production of viral particles that may infect a wide range of mammalian cells (Kost *et al.*, 2005).

F. Cell-Free Expression

Cell-free expression systems are emerging as a promising tool to accelerate the expression of membrane proteins (Cappuccio *et al.*, 2009; Schwarz *et al.*, 2008), and this is reflected by a growing number of HTP reports (Ishihara *et al.*, 2005; Savage *et al.*, 2007). Also, one membrane protein structure has already been determined using protein produced in a cell-free system (Chen *et al.*, 2007). The elimination of a traditional host cell environment can overcome problems associated with inhibitory effects of the recombinant membrane protein or toxicity. Furthermore, these systems offer a range of flexibility in the reaction composition and running formats. For many targets, cell-free strategies have overcome limited expression yields obtained through conventional cell-based methods, including targets that cannot otherwise be expressed to detectable levels (Ishihara *et al.*, 2005; Klammt *et al.*, 2007; Savage *et al.*, 2007). Expression reactions are carried out in small volumes and can be performed within hours, thereby expediting the process of sample production and characterization and eliminating the requirement for large culture rooms and equipment. Prokaryotic and eukaryotic versions of cell-free expression systems have been established (Schwarz *et al.*, 2008) and, in addition to screening solubilizing detergents (Ishihara *et al.*, 2005; Klammt *et al.*, 2007; Savage *et al.*, 2007), factors such as lipids or chaperones can be added directly to the reaction mixture (Kuruma *et al.*, 2005; Schwarz *et al.*, 2007). Cell-free methods have already been utilized for the isotopic labeling of NMR samples (Koglin *et al.*, 2006) and the direct incorporation of heavy atoms for crystallographic phasing (Chen *et al.*, 2007; Savage *et al.*, 2007). Therefore, cell-free approaches can be explored on their own or as a complement to traditional expression systems. Still, batch-to-batch variability, the cost and feasibility of scale-up, improper protein folding, and the lack of posttranslational modifications (depending on the system) are potential limitations that may need to be considered.

G. *Considerations for Monitoring Membrane Protein Expression*

The methods used to monitor protein expression become a major consideration when multiple constructs, vectors, and expression hosts are screened and analyzed in parallel. The choice and location of an affinity tag or fusion protein may dictate the approach used to verify expression, and commercial antibodies are available for most common affinity tags and fusion proteins (Waugh, 2005). Moreover, protein expression *per se* is not necessarily an indicator of the production of *active* protein. As a first pass, one can screen for expression via colony blot (or western blot) of induced whole-cell samples that have been normalized for overall biomass (Lewinson *et al.*, 2008). If a GFP-tag has been employed, whole-cell fluorescence measurements are generally found to correlate with the level of protein expression (Drew *et al.*, 2005; Hammon *et al.*, 2009; Newstead *et al.*, 2007). Subsequent small-scale cellular fractionation or solubilization of expressed samples can help identify the target location (i.e., membrane or insoluble fraction) (Dobrovetsky *et al.*, 2005; Eshaghi *et al.*, 2005; Lewinson *et al.*, 2008; Surade *et al.*, 2006). Once expression and membrane localization of a target has been confirmed, a second wave of expression optimization should be pursued in which different constructs, cell strains, and media conditions are analyzed under various induction protocols. A major key to streamlining an expression screen is to minimize or forgo comprehensive sample processing efforts at an early stage if many negative outcomes are anticipated. Productivity is maximized once preliminary expression data have been obtained, allowing more thorough investigations to be pursued regarding expression optimization and the location and integrity of the target.

Validation extends far beyond the optimization of protein expression levels, although such effort is rarely applied in traditional HTP formats. One example that may be of general utility is the use of confocal microscopy to assess the location of GFP fusion proteins (Eifler *et al.*, 2007; Kawate and Gouaux, 2006; Newstead *et al.*, 2007). When applied to eukaryotic systems, this and immunofluorescence are straightforward methods that can be very fruitful in defining optimal expression constructs by comparing their subcellular location (Rosenbaum *et al.*, 2007). More specialized assays can be applied depending on the target in question. For example, mutant or deletion strains may be used to verify the functional integrity of an expressed target (Lewinson *et al.*, 2009; Ton and Rao, 2004); or in-cell enzymatic and transport assays can be developed. Fluorescent- or radio-labeled substrates can be applied to a wide variety of cell types (or isolated membrane fractions), and this type of ligand-binding assay has been utilized in a HTP format for targeted protein engineering (Sarkar *et al.*, 2008). Patch-clamp electrophysiological techniques can be applied to channel proteins in eukaryotic expression systems (Brock *et al.*, 2001; Eifler *et al.*, 2007; Kawate and Gouaux, 2006; Korepanova *et al.*, 2009) and similar approaches may be possible in bacterial expression systems (Berrier *et al.*, 1989; Cheng *et al.*, 2009; Kuo *et al.*, 2007a, b; Martinac *et al.*, 1987). Although additional effort is required, these types of methodologies may prove extraordinarily productive in assessing and optimizing the stability and functional integrity of diverse target constructs.

Following productive protein expression, obtaining an active, pure, homogenous, and monodisperse sample that is amenable to high-resolution structural studies becomes the next major bottleneck. Validation that extends past the stages of cloning and protein expression is beyond the scope of this review. The reader is advised to review standardized methodologies that have been reported for detergent screening, sample characterization, and optimization (Eshaghi *et al.*, 2005; Hammon *et al.*, 2009; Kawate and Gouaux, 2006; Lewinson *et al.*, 2008; Newby *et al.*, 2009; Newstead *et al.*, 2007; Savage *et al.*, 2007).

III. Considerations for the HTP Expression of Membrane Proteins

A wide variety of HTP membrane protein studies have now been conducted using *E. coli* as the expression host (Bane *et al.*, 2007; Dobrovetsky *et al.*, 2005; Drew *et al.*, 2005; Eshaghi *et al.*, 2005; Hammon *et al.*, 2009; Korepanova *et al.*, 2005; Lewinson *et al.*, 2008; Lundstrom *et al.*, 2006; Michalke *et al.*, 2009; Psakis *et al.*, 2007; Surade *et al.*, 2006). Similar, albeit fewer, investigations have utilized *L. lactis* (Kunji *et al.*, 2003; Monne *et al.*, 2005; Surade *et al.*, 2006), yeast (Andre *et al.*, 2006; Chloupkova *et al.*, 2007; Li *et al.*, 2009; Newstead *et al.*, 2007; White *et al.*, 2007; Yurugi-Kobayashi *et al.*, 2009; Zeder-Lutz *et al.*, 2006), insect, and mammalian cells (Akermoun *et al.*, 2005; Eifler *et al.*, 2007; Hassaine *et al.*, 2006; Lundstrom *et al.*, 2006), as well as cell-free expression systems (Ishihara *et al.*, 2005; Savage *et al.*, 2007; Schwarz *et al.*, 2007). From these HTP studies, three general philosophies have emerged. The classical structural genomics or "shotgun approach" involves the genome-wide cloning of representative members of all membrane protein classes without prior regard for their functional or medical significance. These studies generally aim at finding the "low-hanging fruit" of the membrane proteome and can rapidly identify promising candidates for structural studies, where any new information is typically regarded as biologically significant (Dobrovetsky *et al.*, 2005; Li *et al.*, 2009). Accordingly, the overall number of *de novo* membrane protein structures derived from large-scale structural genomics efforts will continue to grow (Koszelak-Rosenblum *et al.*, 2009; Lunin *et al.*, 2006; Martinez Molina *et al.*, 2007; Page *et al.*, 2009). Similar types of projects, but often of a much smaller scale, also contribute significantly to the increasing number of high-resolution membrane protein structures. Here, two approaches dominate. The "funnel approach" focuses on a specific functional class of protein and involves the cloning of several homologs, orthologs, and paralogs. Expression is generally tested using a variety of different vectors, cell strains, and induction protocols (Lewinson *et al.*, 2008). Typically, multiple constructs produce material suitable for functional studies, and structure determination ultimately provides a template model for the protein class under study (Bass *et al.*, 2002; Chang *et al.*, 1998; Feng *et al.*, 2007; Kawate *et al.*, 2009; Locher *et al.*, 2002; van den Berg *et al.*, 2004). For the structure determination of a single target, the "target-specific approach" may be the most demanding, requiring reiterative testing of different constructs, expression vectors, expression hosts, and expression conditions to achieve suitable levels of functional, homogeneous material (Rosenbaum *et al.*, 2007; Warne *et al.*, 2008).

A technique common to most HTP efforts is the shotgun approach: many genes are cloned into a standard expression vector downstream of an inducible promoter as 6xHis-tag (the most common) or GFP-fusion (less common) constructs. Expression can be screened in a single expression host using a standard induction protocol. In *E. coli*, for example, a standard protocol may employ the C43 strain grown in auto-inducing media with a temperature shift from 37 °C to $\geq$ 22 °C at an OD_{600} of ~0.8 and with growth continued overnight. We note that low-temperature induction in *E. coli* appears to be the most productive (Eshaghi *et al.*, 2005; Lewinson *et al.*, 2008; Wang *et al.*, 2003). Expression is then detected using western or dot blots of whole-cell samples. Small-scale induction cultures can then be further processed with a single detergent for solubilization, purification, and application to an analytical size exclusion column. If the target protein produces a promising chromatographic profile, a large-scale purification is performed, followed by crystallization screens. Remarkably, this basic type of approach has been successful in the crystallization of numerous membrane proteins (Dobrovetsky *et al.*, 2005; Li *et al.*, 2009). To increase expression yields, a more extensive study should include the screening of multiple expression hosts and conditions, as well as the cloning of sequence homologs or multiple constructs of the same target. In general, the likelihood of achieving productive expression levels correlates with the number of variables screened. In *E. coli*, for example, overall rates for prokaryotic targets range from ~30% for single construct and single expression condition trials to ~80% when multiple homologs, vectors, and cell strains are tested (Dobrovetsky *et al.*, 2005; Drew *et al.*, 2005; Eshaghi *et al.*, 2005; Gordon *et al.*, 2008; Hammon *et al.*, 2009; Korepanova *et al.*, 2005; Lewinson *et al.*, 2008; Psakis *et al.*, 2007; Surade *et al.*, 2006). Thus, depending on the specific goals of the research project, such additional pursuits may be employed after protein expression for a given target has been confirmed (or failed).

Clearly, the more rigorous the approach, the more promising the outcomes will be. It is generally advisable to employ a tightly regulated promoter, test amino- and carboxyl-terminal-tagged constructs, with two or more homologous targets, in two or more different cell strains (for

E. coli, at least one being a C41 or C43 derivative). As already mentioned, and discussed below, 6xHis-tag construct expression can be rapidly screened via colony blot analysis, and this can also be coupled downstream to random mutagenesis or other engineering strategies to produce high-expressing variants (e.g., Martinez Molina *et al.*, 2008). Similarly, GFP-tagged constructs have found widespread use in streamlining expression screening, sample characterization, protein engineering, and expression strain evolution strategies (Drew *et al.*, 2005, 2008; Geertsma *et al.*, 2008; Hammon *et al.*, 2009; Ito *et al.*, 2008; Kawate and Gouaux, 2006; Link *et al.*, 2008; Newstead *et al.*, 2007; Skretas and Georgiou, 2008).

IV. Structural Studies of GPCRs

GPCRs feature prominently among drug targets. Over 800 GPCRs are present in the human genome and all share a characteristic arrangement of seven transmembrane α-helices, with the amino-terminus positioned on the extracellular side of the plasma membrane. The binding of ligands such as hormones, neurotransmitters, ions, lipids, or peptides to the extracellular face of a GPCR induces a conformational change that propagates to the intracellular surface, resulting in the activation of G-proteins, and, consequently, changes in the levels of intracellular messengers such as cAMP, Ca^{2+}, and/or signaling lipids. GPCRs are thought to be the targets of up to 50% of drugs currently on the market and 25% of the top 200 best-selling drugs (Lundstrom, 2005; Pierce *et al.*, 2002). Accordingly, drugs targeting GPCRs are indicated in a wide variety of pathologies, including central nervous and immune system disorders, cardiovascular and gastrointestinal diseases, and cancer (Dragun *et al.*, 2009; Li *et al.*, 2005; Lim, 2007). Unfortunately, high-resolution structural information remains limited for the vast majority of GPCRs. As a result, most drugs targeting GPCRs have been identified through classical "ligand-like" screening approaches, without the benefit of structure-based drug design. Structures of bacteriorhodopsin and other prokaryotic homologues have been known for some time, but models of eukaryotic GPCRs based on these templates have limited utility (Okada and Palczewski, 2001). The first high-resolution eukaryotic GPCR structure, that of bovine rhodopsin, significantly advanced our understanding of the structure and mechanism of GPCR activation (Okada and Palczewski, 2001; Palczewski, 2006). But rhodopsin is highly specialized for the detection of light and exhibits functional and biochemical characteristics that distinguish it from other GPCRs. For example, rhodopsin is naturally abundant in retina and, when protected from light, is highly stable in many detergents suitable for structure studies. The natural abundance and detergent stability of most other GPCRs are comparably low (Lundstrom, 2005). Nevertheless, phenomenal progress has recently been achieved in recombinant GPCR expression and structure determination. In this section, we outline the strategies that have been most successful, in addition to highlighting potential limitations or pitfalls of these approaches. It is expected that these strategies will be applicable to the expression of many other eukaryotic membrane proteins.

A. Exploiting E. coli *and Various Selection Techniques for the Expression of GPCRs*

We first examine the use of *E. coli* for GPCR expression since this host remains the most widely used, if not the system of choice, for soluble and membrane protein expression. One key advantage of *E. coli* is its short generation time, as this permits the rapid screening of large numbers of constructs. However, strategies employing *E. coli* also face acute limitations. GPCRs expressed in *E. coli* will not undergo posttranslational modifications such as glycosylation, myristylation, or phosphorylation. While this property can increase the homogeneity of recombinant targets, the folding or activity of some GPCRs may be dependent on posttranslational modifications (Gibson *et al.*, 1998; Kaushal *et al.*, 1994). *E. coli* membranes also lack many lipids normally found in eukaryotic membranes, such as cholesterol and phosphatidylcholine, and this may adversely affect the structural and functional integrity of some recombinant GPCRs (Gimpl *et al.*, 1995, 2008; Hasegawa *et al.*, 1987; Lagane *et al.*, 2000; Nunez and Glass, 1982). Despite these properties, considerable effort has been put forth using *E. coli* for the production of GPCRs, and many GPCRs have been (functionally) expressed. Two approaches dominate. One strategy is to express target GPCRs into inclusion bodies; the incorporation of GPCRs (or other membrane proteins) into intracellular inclusion bodies is not uncommon in *E. coli*. This is followed by purification under denaturing conditions and subsequent protein refolding. One HTP approach has reported the expression of ~40% of all tested GPCRs into inclusion bodies, with yields as high as 350 mg/l (Michalke *et al.*, 2009); a second HTP study has reported comparable results (Lundstrom *et al.*, 2006). However, there are only a few reports of functional GPCR refolding (Baneres *et al.*, 2003, 2005; Kiefer *et al.*, 1996, 1999). The second approach is to express functional GPCRs in the *E. coli* inner membrane. This has proven successful for at least 15 targets, although typically only low-level expression is achieved, usually less than 0.1 mg/l culture or ~50–500 ligand-binding sites/cell (Bertin *et al.*, 1992; Chapot *et al.*, 1990; Furukawa and Haga, 2000; Grisshammer *et al.*, 1993, 1994; Haendler *et al.*, 1993; Hampe *et al.*, 2000; Hulme and Curtis, 1998; Krepkiy *et al.*, 2006; Stanasila *et al.*, 1999; Weiss and Grisshammer, 2002; Xia *et al.*, 1993; Yeliseev *et al.*, 2007). GPCR constructs used for this type of recombinant *E. coli* expression typically employ an amino-terminal fusion that "targets" or directs the expressed protein to the periplasmic space (e.g., MPB) or outer membrane (e.g., LamB; reviewed in(Mancia and Hendrickson, 2007), although other formats are also possible (Hanninen *et al.*, 1994). These strategies can help promote the amino-terminal-out topology of GPCRs. Some successful strategies have also fused Trx to the carboxyl-terminus of the GPCR (e.g., Tucker and Grisshammer, 1996), although the general applicability of these various tagging strategies for the expression of GPCRs in *E. coli* is not obvious (Mancia and Hendrickson, 2007). In some cases, functional ligand binding may require or be enhanced by the addition of exogenous lipids (Gimpl *et al.*, 1995, 2008; Hasegawa *et al.*, 1987; Lagane *et al.*, 2000; Weiss and Grisshammer, 2002). No GPCR structure has currently been solved using material expressed in *E. coli*, but the high-level production of numerous GPCRs has been achieved using various fusion-based approaches (Busuttil *et al.*, 2001; Ren *et al.*, 2009; Yeliseev *et al.*, 2007). Still, *E. coli* may be best suited for the development of efficient screens that identify high-expressing and/or stable constructs, as outlined below.

Christopher Tate and colleagues have recently demonstrated the successful "conformational thermostabilization" of GPCRs by employing a generic protein engineering and expression screen in *E. coli* (Magnani *et al.*, 2008; Serrano-Vega *et al.*, 2008). This has led to the high-resolution structure determination of the turkey β_1-adrenergic receptor in the presence of a short-chain detergent and an antagonist, but in the absence of any fusion protein or antibody fragment (Warne *et al.*, 2008).

Remarkably, the successful construct was first selected using the conformational thermostabilization procedure in *E. coli* and subsequently expressed at high levels in insect cells (Serrano-Vega *et al.*, 2008; Warne *et al.*, 2008). This group has also demonstrated the general utility of their approach on diverse GPCRs expressed in *E. coli*, showing that it can be used to stabilize agonist- as well as antagonist-bound conformations of these receptors (Magnani *et al.*, 2008; Serrano-Vega *et al.*, 2008). The only apparent requirement for this procedure is that some functional GPCR can be expressed in *E. coli*, even at low levels, and that there is some means of measuring receptor activity, such as a simple fluorescent or radioligand-binding assay. Random and/or site-directed mutagenesis techniques are then pursued with the aim of identifying constructs with increased ligand-binding activity or protein stability. Binding and stability assays can be performed with whole-cell samples, crude cell lysates, or detergent-solubilized preparations. Once candidate positions have been identified, stabilizing mutations can be combined to yield GPRC constructs with even more desirable properties. Binding assays may also be repeated in the presence of one or more of the short-chain detergents that are typically used in membrane protein crystallization and/or NMR experiments (Magnani *et al.*, 2008; Serrano-Vega *et al.*, 2008). In this way, candidate constructs can be engineered and selected in *E. coli* and then transferred into expression systems that are typically more robust for eukaryotic membrane protein targets. These second expression hosts may then yield the high-level expression required for structural studies. Importantly, Andreas Plückthun and colleagues have similarly reported the use of a fluorescently labeled GPCR agonist coupled to an error-prone PCR-based method to evolve stable GPCR proteins using *E. coli* expression (Sarkar *et al.*, 2008). It is notable that the evolved GPCRs also expressed at higher levels in all expression systems tested (*E. coli*, *P. pastoris*, and HEK293T cells) relative to the wild-type protein (Sarkar *et al.*, 2008). These types of protein evolution protocols are rapid, can stabilize variant GPCRs in defined conformational states, and may promote the stability of GPCRs in the short-chain detergents that are frequently employed in structural studies. It is clear that these selection methodologies can be extended to other expression systems (Armbruster *et al.*, 2007), which will be critical for GPCRs that fail to express any amount of functional protein in *E. coli*. We note, for example, that one group has demonstrated expression screening of mutant GPCR–GFP proteins in yeast using a rapid mutagenesis-based homologous recombination cloning method (Ito *et al.*, 2008). Still, more specialized or traditional types of engineering strategies have also been employed to stabilize recombinant GPCRs (Martin *et al.*, 2009; Roth *et al.*, 2008; Standfuss *et al.*, 2007).

One of the challenges associated with structural studies of GPCRs is that these receptors have evolved to cycle through distinct conformations, and many GPCRs are known to have intrinsic or constitutive activity without the action of a ligand upon them (Tao, 2008). This is expected to manifest itself as conformational heterogeneity in purified protein samples. The conformational thermostabilization approach of Tate and colleagues (Magnani *et al.*, 2008; Serrano-Vega *et al.*, 2008) and the directed evolution protocols of Sarkar *et al.* (2008) will be of particular utility since both aim to identify stabilizing mutations in the presence of agonist or antagonist-like molecules. It is important to note, however, that there is no *a priori* way to determine which mutations will ultimately have a conformational or thermo-stabilizating effect, even after the structure of the target GPCR is known (Warne *et al.*, 2008). Furthermore, based on the inverse agonist-bound structure of the β_2-adrenergic receptor and subsequent *in silico* structure-based drug screening methods, Brian Kobilka and coworkers have highlighted the likely need for diverse ligand-bound structural repertoires of every GPCR target (Kolb *et al.*, 2009). These facts should serve to reinforce the requirement for HTP and large-scale mutagenesis and screening protocols in GPCR structural biology.

In addition to those already discussed, a collection of other molecular biology techniques should find widespread utility in recombinant GPCR expression. For example, the fusion of GFP to the carboxyl-terminus of a GPCR target has been successfully employed to screen for co-expressed "chaperone" genes or host mutations that ultimately improved recombinant GPCR expression in *E. coli*, where GFP fluorescence provided the basis for selection and cell sorting (Link *et al.*, 2008; Skretas and Georgiou, 2008). Another related strategy has been described that fuses an antibiotic resistance gene to the carboxyl-terminus of the target membrane protein, so that production of the selectable marker and survival on selective media is directly linked to expression (Massey-Gendel *et al.*, 2009). Notably, mutant *E. coli* strains isolated in this manner also showed improved expression for a number of other membrane protein targets that had not undergone the original selection regiment (Massey-Gendel *et al.*, 2009). Further, we note that random mutagenesis protocols have been applied to target constructs in a HTP setting as a general means to increase expression yields of recombinant membrane proteins in *E. coli*, including at least one human target (Martinez Molina *et al.*, 2008). Collectively, these methods should be amenable to HTP techniques in *E. coli* and may also find utility in boosting GPCR expression within eukaryotic expression hosts.

B. Successful Expression Methods for GPCR Structure Determination

The only high-resolution GPCR structures derived from naturally abundant protein sources are those of rhodopsin and its chromophore-free derivative opsin (Murakami and Kouyama, 2008; Palczewski *et al.*, 2000; Park *et al.*, 2008; Scheerer *et al.*, 2008). A high-resolution crystal structure of a recombinant mutant rhodopsin has been obtained after extensive optimization of its expression in mammalian (COS-1) cells (Standfuss *et al.*, 2007), and this extensively developed system has also permitted high-resolution solid-state NMR investigations (Ahuja *et al.*, 2009). Five additional GPCR crystal structures have only recently been determined using recombinant protein material (Cherezov *et al.*, 2007; Hanson *et al.*, 2008; Jaakola *et al.*, 2008; Rasmussen *et al.*, 2007; Warne *et al.*, 2008). Notably, common among all of these latter structures was expression of the target GPCRs in insect cells. These recombinant GPCR targets also underwent significant construct design, as detailed below. Since we have already outlined the promising "conformational thermostabilization" strategies employed by Tate and colleagues leading to the turkey β_1-adrenergic receptor structure in short-chain detergents (Warne *et al.*, 2008), these will not be discussed further here.

Brian Kobilka and coworkers first reported the crystal structure of the β_2-adrenergic receptor in complex with an inverse agonist and antibody fragment (Rasmussen *et al.*, 2007). One key to obtaining this structure was an initial focus on construct design and stability. In fact, all of the recent recombinant GPCR structures have used constructs truncated at the amino- and/or carboxyl-termini in order to remove disordered or protease-liable regions, and in at least one case, sites of posttranslational modification (Warne *et al.*, 2009). Most of these GPCR constructs have also had the third intracellular loop (L3) deleted, as this region is thought to be flexible and therefore prohibitive to productive crystal contact formation (Cherezov *et al.*, 2007; Hanson *et al.*, 2008; Jaakola *et al.*, 2008; Rasmussen *et al.*, 2007; Warne *et al.*, 2008). The first β_2-adrenergic receptor structure was conformationally restricted by the presence of a ligand and an antibody fragment that bound to a discontinuous intracellular epitope. This protein was further stabilized by the use of the bicelle crystallization method (Rasmussen *et al.*, 2007), which likely mimics a membrane environment better than traditional detergent-based crystallization screens (Caffrey, 2003; Faham and Bowie, 2002). An additional β_2-adrenergic receptor structure was solved to higher resolution using a construct in which a soluble protein that is known to crystallize readily, T4 lysozyme, was genetically fused in place of the original L3 loop (Cherezov *et al.*, 2007; Rosenbaum *et al.*, 2007). This work was the first to demonstrate the feasibility of a "carrier protein" approach to promote the crystallization of an integral membrane protein. But, as the authors pointed out, considerable engineering was required to optimize the expression behavior of the final construct (Rosenbaum *et al.*, 2007). Other keys to obtaining this GPCR structure included the inclusion of a ligand (an inverse agonist) to limit conformational flexibility and the use of lipidic cubic phase (LCP) crystallization methods. LCP methods are thought to provide a membrane-like environment that may stabilize and promote the crystallization of some integral membrane proteins (Caffrey, 2003; Caffrey and Cherezov, 2009). Based on these two early structures, Raymond Stevens and coworkers were able to design mutants of the β_2-adrenergic receptor with increased stability, which permitted a shortened purification protocol and higher protein yields (Roth *et al.*, 2008). Ultimately, one mutant β_2-adrenergic receptor-T4-fusion protein crystallized in a new space group to reveal further structural insights (Hanson *et al.*, 2008). In the most recent demonstration of the power of these combined approaches, the Stevens group employed a similar T4-fusion strategy to obtain the high-resolution structure of an antagonist-bound A_{2A}-adenosine receptor crystallized by the LPC method (Jaakola *et al.*, 2008).

V. Considerations for the Expression of Ion Channels

Ion channels represent important targets for the treatment of diverse indications such as chronic pain, cardiovascular, autoimmune, and neurodegenerative diseases. Disruptions in normal ion channel activity have also been linked to cancer, diabetes, and respiratory illnesses (Ashcroft, 2006; Kass, 2005). It is imperative that we obtain high-resolution structural information on ion channel targets, including their complexes with regulatory molecules and pharmacologically relevant drugs. All of the aforementioned techniques discussed for the expression of GPCRs are relevant for the expression of many other eukaryotic membrane proteins, including ion channels. However, eukaryotic ion channels also present some unique expression challenges. Many are extremely large in size (i.e., > 1,000 amino acids) and may contain numerous transmembrane regions. Eukaryotic ion channels also frequently undergo extensive posttranslational modifications and may contain large loops that are sensitive to proteolysis. Below, we briefly review expression strategies that have proven successful for the expression of prokaryotic and eukaryotic ion channel proteins in quantities suitable for structure determination.

The high-resolution structures of several prokaryotic ion channels have significantly advanced our understanding of selectivity, gating, and inactivation (Alam and Jiang, 2009a, b; Bocquet *et al.*, 2009; Cordero-Morales *et al.*, 2006; Hilf and Dutzler, 2009; Jiang *et al.*, 2002; Khademi *et al.*, 2004; Wang *et al.*, 2008; Zhou *et al.*, 2001), as well as other structure–function relationships operating across this diverse protein family. For these targets, the consensus expression strategy is clear: test many homologs for expression in *E. coli*. Douglas Rees and coworkers pioneered this strategy for their work on the MscL and MscS mechanosensitive channels (Bass *et al.*, 2002; Chang *et al.*, 1998; Liu *et al.*, 2009) and have successfully applied it to the structure determination of numerous prokaryotic transporter proteins (Kadaba *et al.*, 2008; Locher *et al.*, 2002; Pinkett *et al.*, 2007). Unfortunately, a consensus expression strategy for eukaryotic ion channels is less clear. There are currently 13 high-resolution recombinant eukaryotic ion channel structures available (Fischer *et al.*, 2009; Gonzales *et al.*, 2009; Hiroaki *et al.*, 2006; Ho *et al.*, 2009; Horsefield *et al.*, 2008; Jasti *et al.*, 2007; Kawate *et al.*, 2009; Long *et al.*, 2005, 2007; Maeda *et al.*, 2009; Newby *et al.*, 2008; Tani *et al.*, 2009; Tornroth-Horsefield *et al.*, 2006) and only three are of human origin (Ho *et al.*, 2009; Horsefield *et al.*, 2008; Maeda *et al.*, 2009). In addition, the expression strategies for these targets are quite varied (e.g., see Table I).

Structures of chicken acid-sensing ion channel 1 (ASIC-1) were achieved following expression in *Sf9* insect cells and implementation of a standardized GFP-screening protocol that included characterization of amino- and carboxyl-terminal-tagged proteins via size exclusion chromatography (Gonzales *et al.*, 2009; Jasti *et al.*, 2007). Notably, a similar approach was employed for the recent zebrafish $P2X_4$ receptor structure (Kawate *et al.*, 2009). Structures of the human connexin 26 gap junction channel and of rat aquaporin-4 have also been solved following their expression in *Sf9* insect cells (Hiroaki *et al.*, 2006; Maeda *et al.*, 2009; Tani *et al.*, 2009). However, both of the available rat Kv1.2 channel structures have been obtained following yeast cell expression (Long *et al.*, 2005, 2007). Similarly, human aquaporin-4, human aquaporin-5, and plant aquaporin SoPIP2;1 structures have also derived from *P. pastoris* expressed material (Ho *et al.*, 2009; Horsefield *et al.*, 2008; Tornroth-Horsefield *et al.*, 2006).

Perhaps one rule of thumb when pursuing the expression of eukaryotic ion channels is to first survey the expression results for the same or related channels that are already present in the literature. For example, high-resolution crystal structures of aquaporin-4, aquaporin-5, and spinach aquaporin SoPIP2;1 were determined following their recombinant expression in yeast (Ho *et al.*, 2009; Horsefield *et al.*, 2008; Tornroth-Horsefield *et al.*, 2006), and an intermediate resolution crystal structure of recombinant bovine aquaporin-0 has also been derived this way (Palanivelu *et al.*, 2006). Therefore, yeast expression may be indicated when pursing this class of eukaryotic channel; but it is also noteworthy that a high-resolution structure of aquaglyceroporin from *Plasmodium falciparum* has recently been obtained following recombinant expression in *E. coli* (Newby *et al.*, 2008). Therefore, it is likely that standard cloning, expression, and screening methods like those outlined in the former section on GPCRs will continue to give rise to material suitable for high-resolution structure determination. The design and/or selection of suitable constructs will be the critical first step in achieving high-level expression of stable, homogeneous protein. We suggest a rational approach in which the number of constructs is initially kept to a minimum, thereby ensuring that expression screening and detergent solubilization experiments do not become early bottlenecks. We consider the approaches pioneered by Eric Gouaux and colleagues as benchmarks, where these involve the screening of many homologs with both amino- and carboxyl-terminal tagging strategies (Gonzales *et al.*, 2009; Jasti *et al.*,

2007; Kawate and Gouaux, 2006; Kawate *et al.*, 2009). However, there is no *a priori* way to know which construct or host will perform best. For instance, different research groups have concluded that amino- or carboxyl-terminally tagged constructs of P2X receptor channels resulted in higher yields when expressed in insect (*Sf9*) or mammalian (HEK) cells, respectively (Kawate and Gouaux, 2006; Mio *et al.*, 2009).

VI. Conclusions

It is clear that obtaining high-level expression of recombinant membrane proteins for structural studies will continue to be challenging. While the ultimate guide would be to "try everything," this is clearly impossible in most laboratory settings. A reasonable compromise is to follow the common, successful trends. First, *E. coli* is the expression system of choice for prokaryotic membrane targets and should also be considered a good starting point for many, if not most, eukaryotic membrane proteins. Second, failing expression in *E. coli*, eukaryotic targets should be tested in *both* yeast and insect cell expression systems; because, as yet, there is no clear front-runner between the two. Third, multiple constructs and fusion strategies should always be attempted, regardless of the expression host. Fourth, methods to evolve conformationally stabilized or highly expressing mutant proteins (or hosts) can help drive structure determination for the most challenging or intractable targets, particularly those of eukaryotic origin. Lastly, emerging technologies, such as cell-free expression, are likely to help yield important new insights into membrane protein structure and function.

References

Ago, H., Kanaoka, Y., Irikura, D., Lam, B. K., Shimamura, T., Austen, K. F., and Miyano, M. (2007). Crystal structure of a human membrane protein involved in cysteinyl leukotriene biosynthesis. *Nature* **448**, 609–612.

Ahuja, S., Hornak, V., Yan, E. C., Syrett, N., Goncalves, J. A., Hirshfeld, A., Ziliox, M., Sakmar, T. P., Sheves, M., Reeves, P. J., Smith, S. O., and Eilers, M. (2009). Helix movement is coupled to displacement of the second extracellular loop in rhodopsin activation. *Nat. Struct. Mol. Biol.* **16**, 168–175.

Akermoun, M., Koglin, M., Zvalova-Iooss, D., Folschweiller, N., Dowell, S. J., and Gearing, K. L. (2005). Characterization of 16 human G protein-coupled receptors expressed in baculovirus-infected insect cells. *Protein Expr. Purif.* **44**, 65–74.

Alam, A., and Jiang, Y. (2009a). High-resolution structure of the open NaK channel. *Nat. Struct. Mol. Biol.* **16**, 30–34.

Alam, A., and Jiang, Y. (2009b). Structural analysis of ion selectivity in the NaK channel. *Nat. Struct. Mol. Biol.* **16**, 35–41.

Andre, N., Cherouati, N., Prual, C., Steffan, T., Zeder-Lutz, G., Magnin, T., Pattus, F., Michel, H., Wagner, R., and Reinhart, C. (2006). Enhancing functional production of G protein-coupled receptors in *Pichia pastoris* to levels required for structural studies via a single expression screen. *Protein Sci.* **15**, 1115–1126.

Armbruster, B. N., Li, X., Pausch, M. H., Herlitze, S., and Roth, B. L. (2007). Evolving the lock to fit the key to create a family of G protein-coupled receptors potently activated by an inert ligand. *Proc. Natl. Acad. Sci. USA* **104**, 5163–5168.

Ashcroft, F. M. (2006). From molecule to malady. *Nature* **440**, 440–447.

Attrill, H., Harding, P. J., Smith, E., Ross, S., and Watts, A. (2009). Improved yield of a ligand-binding GPCR expressed in *E. coli* for structural studies. *Protein Expr. Purif.* **64**, 32–38.

Auer, M., Kim, M. J., Lemieux, M. J., Villa, A., Song, J., Li, X. D., and Wang, D. N. (2001). High-yield expression and functional analysis of *Escherichia coli* glycerol-3-phosphate transporter. *Biochemistry* **40**, 6628–6635.

Bane, S. E., Velasquez, J. E., and Robinson, A. S. (2007). Expression and purification of milligram levels of inactive G-protein coupled receptors in *E. coli*. *Protein Expr. Purif.* **52**, 348–355.

Baneres, J. L., Martin, A., Hullot, P., Girard, J. P., Rossi, J. C., and Parello, J. (2003). Structure-based analysis of GPCR function: Conformational adaptation of both agonist and receptor upon leukotriene B4 binding to recombinant BLT1. *J. Mol. Biol.* **329**, 801–814.

Baneres, J. L., Mesnier, D., Martin, A., Joubert, L., Dumuis, A., and Bockaert, J. (2005). Molecular characterization of a purified 5-HT4 receptor: A structural basis for drug efficacy. *J. Biol. Chem.* **280**, 20253–20260.

Bass, R. B., Strop, P., Barclay, M., and Rees, D. C. (2002). Crystal structure of *Escherichia coli* MscS, a voltage-modulated and mechanosensitive channel. *Science* **298**, 1582–1587.

Berntsson, R. P., Alia Oktaviani, N., Fusetti, F., Thunnissen, A. M., Poolman, B., and Slotboom, D. J. (2009). Selenomethionine incorporation in proteins expressed in *Lactococcus lactis*. *Protein Sci.* **18**, 1121–1127.

Berrier, C., Coulombe, A., Houssin, C., and Ghazi, A. (1989). A patch-clamp study of ion channels of inner and outer membranes and of contact zones of *E. coli*, fused into giant liposomes. Pressure-activated channels are localized in the inner membrane. *FEBS Lett.* **259**, 27–32.

Berrow, N. S., Alderton, D., Sainsbury, S., Nettleship, J., Assenberg, R., Rahman, N., Stuart, D. I., and Owens, R. J. (2007). A versatile ligation-independent cloning method suitable for high-throughput expression screening applications. *Nucleic Acids Res.* **35**, e45.

Bertin, B., Freissmuth, M., Breyer, R. M., Schutz, W., Strosberg, A. D., and Marullo, S. (1992). Functional expression of the human serotonin 5-HT1A receptor in

Escherichia coli. Ligand binding properties and interaction with recombinant G protein alpha-subunits. *J. Biol. Chem.* **267**, 8200–8206.

Blasey, H. D., Brethon, B., Hovius, R., Vogel, H. H., Tairi, A. P., Lundstrom, K., Rey, L., and Bernard, A. R. (2000). Large scale transient 5-HT3 receptor production with the Semliki Forest Virus Expression System. *Cytotechnology* **32**, 199–208.

Blundell, T. L., Elliott, G., Gardner, S. P., Hubbard, T., Islam, S., Johnson, M., Mantafounis, D., Murray-Rust, P., Overington, J., Pitts, J. E., et al., (1989). Protein engineering and design. *Philos. Trans. R. Soc. Lond. B Biol. Sci.* **324**, 447–460.

Bocquet, N., Nury, H., Baaden, M., Le Poupon, C., Changeux, J. P., Delarue, M., and Corringer, P. J. (2009). X-ray structure of a pentameric ligand-gated ion channel in an apparently open conformation. *Nature* **457**, 111–114.

Brinkmann, U., Mattes, R. E., and Buckel, P. (1989). High-level expression of recombinant genes in *Escherichia coli* is dependent on the availability of the *dnaY* gene product. *Gene* **85**, 109–114.

Brock, M. W., Lebaric, Z. N., Neumeister, H., DeTomaso, A., and Gilly, W. F. (2001). Temperature-dependent expression of a squid Kv1 channel in *Sf9* cells and functional comparison with the native delayed rectifier. *J. Membr. Biol.* **180**, 147–161.

Busuttil, B. E., Turney, K. L., and Frauman, A. G. (2001). The expression of soluble, full-length, recombinant human TSH receptor in a prokaryotic system. *Protein Expr. Purif.* **23**, 369–373.

Cacquevel, M., Aeschbach, L., Osenkowski, P., Li, D., Ye, W., Wolfe, M. S., Li, H., Selkoe, D. J., and Fraering, P. C. (2008). Rapid purification of active gamma-secretase, an intramembrane protease implicated in Alzheimer's disease. *J. Neurochem.* **104**, 210–220.

Caffrey, M. (2003). Membrane protein crystallization. *J. Struct. Biol.* **142**, 108–132.

Caffrey, M., and Cherezov, V. (2009). Crystallizing membrane proteins using lipidic mesophases. *Nat. Protoc.* **4**, 706–731.

Call, M. E., Schnell, J. R., Xu, C., Lutz, R. A., Chou, J. J., and Wucherpfennig, K. W. (2006). The structure of the zetazeta transmembrane dimer reveals features essential for its assembly with the T cell receptor. *Cell* **127**, 355–368.

Cappuccio, J. A., Hinz, A. K., Kuhn, E. A., Fletcher, J. E., Arroyo, E. S., Henderson, P. T., Blanchette, C. D., Walsworth, V. L., Corzett, M. H., Law, R. J., Pesavento, J. B., Segelke, B. W., Sulchek, T. A., Chromy, B. A., Katzen, F., Peterson, T., Bench, G., Kudlicki, W., Hoeprich, P. D., Jr., and Coleman, M. A. (2009). Cell-free expression for nanolipoprotein particles: Building a high-throughput membrane protein solubility platform. *Methods Mol. Biol.* **498**, 273–296.

Chang, G., Spencer, R. H., Lee, A. T., Barclay, M. T., and Rees, D. C. (1998). Structure of the MscL homolog from *Mycobacterium tuberculosis*: A gated mechanosensitive ion channel. *Science* **282**, 2220–2226.

Chapot, M. P., Eshdat, Y., Marullo, S., Guillet, J. G., Charbit, A., Strosberg, A. D., Delavier-Klutchko, C. (1990). Localization and characterization of three different beta-adrenergic receptors expressed in *Escherichia coli*. *Eur. J. Biochem.* **187**, 137–144.

Chen, Y. J., Pornillos, O., Lieu, S., Ma, C., Chen, A. P., and Chang, G. (2007). X-ray structure of EmrE supports dual topology model. *Proc. Natl. Acad. Sci. USA* **104**, 18999–19004.

Cheng, W. W., Enkvetchakul, D., and Nichols, C. G. (2009). KirBac1.1: It's an inward rectifying potassium channel. *J. Gen. Physiol.* **133**, 295–305.

Cherezov, V., Rosenbaum, D. M., Hanson, M. A., Rasmussen, S. G., Thian, F. S., Kobilka, T. S., Choi, H. J., Kuhn, P., Weis, W. I., Kobilka, B. K., and Stevens, R. C. (2007). High-resolution crystal structure of an engineered human beta2-adrenergic G protein-coupled receptor. *Science* **318**, 1258–1265.

Chloupkova, M., Pickert, A., Lee, J. Y., Souza, S., Trinh, Y. T., Connelly, S. M., Dumont, M. E., Dean, M., and Urbatsch, I. L. (2007). Expression of 25 human ABC transporters in the yeast *Pichia pastoris* and characterization of the purified ABCC3 ATPase activity. *Biochemistry* **46**, 7992–8003.

Cordero-Morales, J. F., Cuello, L. G., Zhao, Y., Jogini, V., Cortes, D. M., Roux, B., and Perozo, E. (2006). Molecular determinants of gating at the potassium-channel selectivity filter. *Nat. Struct. Mol. Biol.* **13**, 311–318.

Cowan, S. W., Schirmer, T., Rummel, G., Steiert, M., Ghosh, R., Pauptit, R. A., Jansonius, J. N., and Rosenbusch, J. P. (1992). Crystal structures explain functional properties of two *E. coli* porins. *Nature* **358**, 727–733.

Deacon, S. E., Roach, P. C., Postis, V. L., Wright, G. S., Xia, X., Phillips, S. E., Knox, J. P., Henderson, P. J., McPherson, M. J., and Baldwin, S. A. (2008). Reliable scale-up of membrane protein over-expression by bacterial auto-induction: From microwell plates to pilot scale fermentations. *Mol. Membr. Biol.* **25**, 588–598.

Deisenhofer, J., Epp, O., Miki, K., Huber, R., and Michel, H. (1985). Structure of the protein subunits in the photosynthetic reaction centre of *Rhodopseudomonas viridis* at 3 Å resolution. *Nature* **318**, 618–624.

Del Tito, B. J., Jr., Ward, J. M., Hodgson, J., Gershater, C. J., Edwards, H., Wysocki, L. A., Watson, F. A., Sathe, G., and Kane, J. F. (1995). Effects of a minor isoleucyl tRNA on heterologous protein translation in *Escherichia coli*. *J. Bacteriol.* **177**, 7086–7091.

Dobrovetsky, E., Lu, M. L., Andorn-Broza, R., Khutoreskaya, G., Bray, J. E., Savchenko, A., Arrowsmith, C. H., Edwards, A. M., and Koth, C. M. (2005). High-throughput production of prokaryotic membrane proteins. *J. Struct. Funct. Genomics* **6**, 33–50.

Doyle, D. A., Morais Cabral, J., Pfuetzner, R. A., Kuo, A., Gulbis, J. M., Cohen, S. L., Chait, B. T., and MacKinnon, R. (1998). The structure of the potassium channel: Molecular basis of K^+ conduction and selectivity. *Science* **280**, 69–77.

Dragun, D., Philippe, A., Catar, R., and Hegner, B. (2009). Autoimmune mediated G-protein receptor activation in cardiovascular and renal pathologies. Thromb Haemost. **101**, 643–648.

Drew, D., Newstead, S., Sonoda, Y., Kim, H., von Heijne, G., and Iwata, S. (2008). GFP-based optimization scheme for the overexpression and purification of eukaryotic membrane proteins in *Saccharomyces cerevisiae*. *Nat. Protoc.* **3**, 784–798.

Drew, D., Slotboom, D. J., Friso, G., Reda, T., Genevaux, P., Rapp, M., Meindl-Beinker, N. M., Lambert, W., Lerch, M., Daley, D. O., Van Wijk, K. J., Hirst, J.,

Kunji, E., De Gier, J. W. (2005). A scalable, GFP-based pipeline for membrane protein overexpression screening and purification. *Protein Sci.* **14**, 2011–2017.

Eifler, N., Duckely, M., Sumanovski, L. T., Egan, T. M., Oksche, A., Konopka, J. B., Luthi, A., Engel, A., and Werten, P. J. (2007). Functional expression of mammalian receptors and membrane channels in different cells. *J. Struct. Biol.* **159**, 179–193.

Eshaghi, S., Hedren, M., Nasser, M. I., Hammarberg, T., Thornell, A., and Nordlund, P. (2005). An efficient strategy for high-throughput expression screening of recombinant integral membrane proteins. *Protein Sci.* **14**, 676–683.

Faham, S., and Bowie, J. U. (2002). Bicelle crystallization: A new method for crystallizing membrane proteins yields a monomeric bacteriorhodopsin structure. *J. Mol. Biol.* **316**, 1–6.

Faller, M., Niederweis, M., and Schulz, G. E. (2004). The structure of a mycobacterial outer-membrane channel. *Science* **303**, 1189–1192.

Feng, L., Yan, H., Wu, Z., Yan, N., Wang, Z., Jeffrey, P. D., and Shi, Y. (2007). Structure of a site-2 protease family intramembrane metalloprotease. *Science* **318**, 1608–1612.

Fischer, G., Kosinska-Eriksson, U., Aponte-Santamaria, C., Palmgren, M., Geijer, C., Hedfalk, K., Hohmann, S., de Groot, B. L., Neutze, R., Lindkvist-Petersson, K. (2009). Crystal structure of a yeast aquaporin at 1.15 ängstrom reveals a novel gating mechanism. *PLoS Biol.* **7**, e1000130.

Fu, C., Wehr, D. R., Edwards, J., and Hauge, B. (2008). Rapid one-step recombinational cloning. *Nucleic Acids Res.* **36**, e54.

Furukawa, H., and Haga, T. (2000). Expression of functional M2 muscarinic acetylcholine receptor in *Escherichia coli*. *J. Biochem.* **127**, 151–161.

Geertsma, E. R., Groeneveld, M., Slotboom, D. J., and Poolman, B. (2008). Quality control of overexpressed membrane proteins. *Proc. Natl. Acad. Sci. USA* **105**, 5722–5727.

Gibson, S. K., Parkes, J. H., and Liebman, P. A. (1998). Phosphorylation stabilizes the active conformation of rhodopsin. *Biochemistry* **37**, 11393–11398.

Gimpl, G., Klein, U., Reilander, H., and Fahrenholz, F. (1995). Expression of the human oxytocin receptor in baculovirus-infected insect cells: High-affinity binding is induced by a cholesterol-cyclodextrin complex. *Biochemistry* **34**, 13794–13801.

Gimpl, G., Reitz, J., Brauer, S., and Trossen, C. (2008). Oxytocin receptors: Ligand binding, signalling and cholesterol dependence. *Prog. Brain Res.* **170**, 193–204.

Gonzales, E. B., Kawate, T., and Gouaux, E. (2009). Pore architecture and ion sites in acid-sensing ion channels and P2X receptors. *Nature* **460**, 599–604.

Gordon, E., Horsefield, R., Swarts, H. G., de Pont, J. J., Neutze, R., and Snijder, A. (2008). Effective high-throughput overproduction of membrane proteins in *Escherichia coli*. *Protein Expr. Purif.* **62**, 1–8.

Grisshammer, R. (2006). Understanding recombinant expression of membrane proteins. *Curr. Opin. Biotechnol.* **17**, 337–340.

Grisshammer, R., Duckworth, R., and Henderson, R. (1993). Expression of a rat neurotensin receptor in *Escherichia coli*. *Biochem. J.* **295**(Pt 2), 571–576.

Grisshammer, R., Little, J., and Aharony, D. (1994). Expression of rat NK-2 (neurokinin A) receptor in *E. coli*. *Receptors Channels* **2**, 295–302.

Guzman, L. M., Belin, D., Carson, M. J., and Beckwith, J. (1995). Tight regulation, modulation, and high-level expression by vectors containing the arabinose PBAD promoter. *J. Bacteriol.* **177**, 4121–4130.

Haendler, B., Hechler, U., Becker, A., and Schleuning, W. D. (1993). Expression of human endothelin receptor ETB by *Escherichia coli* transformants. *Biochem. Biophys. Res. Commun.* **191**, 633–638.

Haldankar, R., Li, D., Saremi, Z., Baikalov, C., and Deshpande, R. (2006). Serum-free suspension large-scale transient transfection of CHO cells in WAVE bioreactors. *Mol. Biotechnol.* **34**, 191–199.

Hamilton, S. R., and Gerngross, T. U. (2007). Glycosylation engineering in yeast: The advent of fully humanized yeast. *Curr. Opin. Biotechnol.* **18**, 387–392.

Hammon, J., Palanivelu, D. V., Chen, J., Patel, C., Minor, D. L., Jr. (2009). A green fluorescent protein screen for identification of well-expressed membrane proteins from a cohort of extremophilic organisms. *Protein Sci.* **18**, 121–133.

Hampe, W., Voss, R. H., Haase, W., Boege, F., Michel, H., and Reilander, H. (2000). Engineering of a proteolytically stable human beta 2-adrenergic receptor/maltose-binding protein fusion and production of the chimeric protein in *Escherichia coli* and baculovirus-infected insect cells. *J. Biotechnol.* **77**, 219–234.

Hanninen, A. L., Bamford, D. H., and Grisshammer, R. (1994). Expression in *Escherichia coli* of rat neurotensin receptor fused to membrane proteins from the membrane-containing bacteriophage PRD1. *Biol. Chem. Hoppe Seyler* **375**, 833–836.

Hanson, M. A., Cherezov, V., Griffith, M. T., Roth, C. B., Jaakola, V. P., Chien, E. Y., Velasquez, J., Kuhn, P., and Stevens, R. C. (2008). A specific cholesterol binding site is established by the 2.8 Å structure of the human beta2-adrenergic receptor. *Structure* **16**, 897–905.

Hartley, J. L., Temple, G. F., and Brasch, M. A. (2000). DNA cloning using *in vitro* site-specific recombination. *Genome Res.* **10**, 1788–1795.

Hasegawa, J., Loh, H. H., and Lee, N. M. (1987). Lipid requirement for mu opioid receptor binding. *J. Neurochem.* **49**, 1007–1012.

Hassaine, G., Wagner, R., Kempf, J., Cherouati, N., Hassaine, N., Prual, C., Andre, N., Reinhart, C., Pattus, F., and Lundstrom, K. (2006). Semliki Forest virus vectors for overexpression of 101 G protein-coupled receptors in mammalian host cells. *Protein Expr. Purif.*. **45**, 343–351.

Hattori, M., Tanaka, Y., Fukai, S., Ishitani, R., and Nureki, O. (2007). Crystal structure of the MgtE Mg^{2+} transporter. *Nature* **448**, 1072–1075.

Higgins, M. K., Demir, M., and Tate, C. G. (2003). Calnexin co-expression and the use of weaker promoters increase the expression of correctly assembled Shaker potassium channel in insect cells. *Biochim. Biophys. Acta* **1610**, 124–132.

Hilf, R. J., and Dutzler, R. (2008). X-ray structure of a prokaryotic pentameric ligand-gated ion channel. *Nature* **452**, 375–379.

Hilf, R. J., and Dutzler, R. (2009). Structure of a potentially open state of a proton-activated pentameric ligand-gated ion channel. *Nature* **457**, 115–118.

Hiroaki, Y., Tani, K., Kamegawa, A., Gyobu, N., Nishikawa, K., Suzuki, H., Walz, T., Sasaki, S., Mitsuoka, K., Kimura, K., Mizoguchi, A., and Fujiyoshi, Y. (2006). Implications of the aquaporin-4 structure on array formation and cell adhesion. *J. Mol. Biol.* **355**, 628–639.

Ho, J. D., Yeh, R., Sandstrom, A., Chorny, I., Harries, W. E., Robbins, R. A., Miercke, L. J., and Stroud, R. M. (2009). Crystal structure of human aquaporin 4 at 1.8 Å and its mechanism of conductance. *Proc. Natl. Acad. Sci. USA* **106**, 7437–7442.

Ho, Y., Lo, H. R., Lee, T. C., Wu, C. P., and Chao, Y. C. (2004). Enhancement of correct protein folding *in vivo* by a non-lytic baculovirus. *Biochem. J.* **382**, 695–702.

Horsefield, R., Norden, K., Fellert, M., Backmark, A., Tornroth-Horsefield, S., Terwisscha van Scheltinga, A. C., Kvassman, J., Kjellbom, P., Johanson, U., and Neutze, R. (2008). High-resolution x-ray structure of human aquaporin 5. *Proc. Natl. Acad. Sci. USA* **105**, 13327–13332.

Hu, J., Qin, H., Li, C., Sharma, M., Cross, T. A., and Gao, F. P. (2007). Structural biology of transmembrane domains: Efficient production and characterization of transmembrane peptides by NMR. *Protein Sci.* **16**, 2153–2165.

Hulme, E. C., and Curtis, C. A. (1998). Purification of recombinant M1 muscarinic acetylcholine receptor. *Biochem. Soc. Trans.* **26**, S361.

Ishihara, G., Goto, M., Saeki, M., Ito, K., Hori, T., Kigawa, T., Shirouzu, M., and Yokoyama, S. (2005). Expression of G protein coupled receptors in a cell-free translational system using detergents and thioredoxin-fusion vectors. *Protein Expr. Purif.* **41**, 27–37.

Ito, K., Sugawara, T., Shiroishi, M., Tokuda, N., Kurokawa, A., Misaka, T., Makyio, H., Yurugi-Kobayashi, T., Shimamura, T., Nomura, N., Murata, T., Abe, K., Iwata, S., and Kobayashi, T. (2008). Advanced method for high-throughput expression of mutated eukaryotic membrane proteins in *Saccharomyces cerevisiae*. *Biochem. Biophys. Res. Commun.* **371**, 841–845.

Jaakola, V. P., Griffith, M. T., Hanson, M. A., Cherezov, V., Chien, E. Y., Lane, J. R., Ijzerman, A. P., and Stevens, R. C. (2008). The 2.6 angstrom crystal structure of a human A2A adenosine receptor bound to an antagonist. *Science* **322**, 1211–1217.

Jana, S., and Deb, J. K. (2005). Strategies for efficient production of heterologous proteins in *Escherichia coli*. *Appl. Microbiol. Biotechnol.* **67**, 289–298.

Jasti, J., Furukawa, H., Gonzales, E. B., and Gouaux, E. (2007). Structure of acid-sensing ion channel 1 at 1.9 Å resolution and low pH. *Nature* **449**, 316–323.

Jiang, Y., Lee, A., Chen, J., Cadene, M., Chait, B. T., and MacKinnon, R. (2002). Crystal structure and mechanism of a calcium-gated potassium channel. *Nature* **417**, 515–522.

Jiang, Y., Lee, A., Chen, J., Ruta, V., Cadene, M., Chait, B. T., and MacKinnon, R. (2003). X-ray structure of a voltage-dependent K^+ channel. *Nature* **423**, 33–41.

Jidenko, M., Nielsen, R. C., Sorensen, T. L., Moller, J. V., le Maire, M., Nissen, P., and Jaxel, C. (2005). Crystallization of a mammalian membrane protein overexpressed in *Saccharomyces cerevisiae*. *Proc. Natl. Acad. Sci. USA* **102**, 11687–11691.

Junge, F., Schneider, B., Reckel, S., Schwarz, D., Dotsch, V., and Bernhard, F. (2008). Large-scale production of functional membrane proteins. *Cell Mol. Life Sci.* **65**, 1729–1755.

Kadaba, N. S., Kaiser, J. T., Johnson, E., Lee, A., and Rees, D. C. (2008). The high-affinity *E. coli* methionine ABC transporter: Structure and allosteric regulation. *Science* **321**, 250–253.

Kass, R. S. (2005). The channelopathies: Novel insights into molecular and genetic mechanisms of human disease. *J. Clin. Invest.* **115**, 1986–1989.

Kaushal, S., Ridge, K. D., and Khorana, H. G. (1994). Structure and function in rhodopsin: The role of asparagine-linked glycosylation. *Proc. Natl. Acad. Sci. USA* **91**, 4024–4028.

Kawate, T., and Gouaux, E. (2006). Fluorescence-detection size-exclusion chromatography for precrystallization screening of integral membrane proteins. *Structure* **14**, 673–681.

Kawate, T., Michel, J. C., Birdsong, W. T., and Gouaux, E. (2009). Crystal structure of the ATP-gated P2X(4) ion channel in the closed state. *Nature* **460**, 592–598.

Khademi, S., O'Connell, J., 3rd, Remis, J., Robles-Colmenares, Y., Miercke, L. J., and Stroud, R. M. (2004). Mechanism of ammonia transport by Amt/MEP/Rh: Structure of AmtB at 1.35 Å. *Science* **305**, 1587–1594.

Kiefer, H., Krieger, J., Olszewski, J. D., Von Heijne, G., Prestwich, G. D., and Breer, H. (1996). Expression of an olfactory receptor in *Escherichia coli*: Purification, reconstitution, and ligand binding. *Biochemistry* **35**, 16077–16084.

Kiefer, H., Maier, K., and Vogel, R. (1999). Refolding of G-protein-coupled receptors from inclusion bodies produced in *Escherichia coli. Biochem. Soc. Trans.* **27**, 908–912.

Klammt, C., Schwarz, D., Eifler, N., Engel, A., Piehler, J., Haase, W., Hahn, S., Dotsch, V., and Bernhard, F. (2007). Cell-free production of G protein-coupled receptors for functional and structural studies. *J. Struct. Biol.* **158**, 482–493.

Koglin, A., Klammt, C., Trbovic, N., Schwarz, D., Schneider, B., Schafer, B., Lohr, F., Bernhard, F., and Dotsch, V. (2006). Combination of cell-free expression and NMR spectroscopy as a new approach for structural investigation of membrane proteins. *Magn. Reson. Chem.* **44** Spec No, S17–S23.

Kolb, P., Rosenbaum, D. M., Irwin, J. J., Fung, J. J., Kobilka, B. K., and Shoichet, B. K. (2009). Structure-based discovery of beta2-adrenergic receptor ligands. *Proc. Natl. Acad. Sci. USA* **106**, 6843–6848.

Korepanova, A., Gao, F. P., Hua, Y., Qin, H., Nakamoto, R. K., and Cross, T. A. (2005). Cloning and expression of multiple integral membrane proteins from *Mycobacterium tuberculosis* in *Escherichia coli. Protein Sci.* **14**, 148–158.

Korepanova, A., Moore, J. D., Nguyen, H. B., Hua, Y., Cross, T. A., and Gao, F. (2007). Expression of membrane proteins from *Mycobacterium tuberculosis* in *Escherichia coli* as fusions with maltose binding protein. *Protein Expr. Purif.* **53**, 24–30.

Korepanova, A., Pereda-Lopez, A., Solomon, L. R., Walter, K. A., Lake, M. R., Bianchi, B. R., McDonald, H. A., Neelands, T. R., Shen, J., Matayoshi, E. D., Moreland, R. B., and Chiu, M. L. (2009). Expression and purification of human TRPV1 in baculovirus-infected insect cells for structural studies. *Protein Expr. Purif.* **65**, 38–50.

Kost, T. A., Condreay, J. P., and Jarvis, D. L. (2005). Baculovirus as versatile vectors for protein expression in insect and mammalian cells. *Nat. Biotechnol.* **23**, 567–575.

Koszelak-Rosenblum, M., Krol, A., Mozumdar, N., Wunsch, K., Ferin, A., Cook, E., Veatch, C. K., Nagel, R., Luft, J. R., Detitta, G. T., and Malkowski, M. G. (2009).

Determination and application of empirically derived detergent phase boundaries to effectively crystallize membrane proteins. *Protein Sci.* **18**, 1828–1839.

Krautwurst, D., Yau, K. W., and Reed, R. R. (1998). Identification of ligands for olfactory receptors by functional expression of a receptor library. *Cell* **95**, 917–926.

Krepkiy, D., Wong, K., Gawrisch, K., and Yeliseev, A. (2006). Bacterial expression of functional, biotinylated peripheral cannabinoid receptor CB2. *Protein Expr. Purif.* **49**, 60–70.

Kudla, G., Murray, A. W., Tollervey, D., and Plotkin, J. B. (2009). Coding-sequence determinants of gene expression in *Escherichia coli*. *Science* **324**, 255–258.

Kunji, E. R., Slotboom, D. J., and Poolman, B. (2003). *Lactococcus lactis* as host for overproduction of functional membrane proteins. *Biochim. Biophys. Acta* **1610**, 97–108.

Kuo, M. M., Baker, K. A., Wong, L., Choe, S. (2007a). Dynamic oligomeric conversions of the cytoplasmic RCK domains mediate MthK potassium channel activity. *Proc. Natl. Acad. Sci. USA* **104**, 2151–2156.

Kuo, M. M., Saimi, Y., Kung, C., Choe, S. (2007b). Patch clamp and phenotypic analyses of a prokaryotic cyclic nucleotide-gated K^+ channel using *Escherichia coli* as a host. *J. Biol. Chem.* **282**, 24294–24301.

Kuruma, Y., Nishiyama, K., Shimizu, Y., Muller, M., and Ueda, T. (2005). Development of a minimal cell-free translation system for the synthesis of presecretory and integral membrane proteins. *Biotechnol. Prog.* **21**, 1243–1251.

Lagane, B., Gaibelet, G., Meilhoc, E., Masson, J. M., Cezanne, L., and Lopez, A. (2000). Role of sterols in modulating the human mu-opioid receptor function in *Saccharomyces cerevisiae*. *J. Biol. Chem.* **275**, 33197–33200.

Lewinson, O., Lee, A. T., and Rees, D. C. (2008). The funnel approach to the precrystallization production of membrane proteins. *J. Mol. Biol.* **377**, 62–73.

Lewinson, O., Lee, A. T., and Rees, D. C. (2009). A P-type ATPase importer that discriminates between essential and toxic transition metals. *Proc. Natl. Acad. Sci. USA* **106**, 4677–4682.

Li, M., Hays, F. A., Roe-Zurz, Z., Vuong, L., Kelly, L., Ho, C. M., Robbins, R. M., Pieper, U., O'Connell, J. D., 3rd, Miercke, L. J., Giacomini, K. M., Sali, A., and Stroud, R. M. (2009). Selecting optimum eukaryotic integral membrane proteins for structure determination by rapid expression and solubilization screening. *J. Mol. Biol.* **385**, 820–830.

Li, S., Huang, S., and Peng, S. B. (2005). Overexpression of G protein-coupled receptors in cancer cells: Involvement in tumor progression. *Int. J. Oncol.* **27**, 1329–1339.

Lim, W. K. (2007). GPCR drug discovery: Novel ligands for CNS receptors. Recent Pat CNS *Drug Discov.* **2**, 107–112.

Link, A. J., Skretas, G., Strauch, E. M., Chari, N. S., and Georgiou, G. (2008). Efficient production of membrane-integrated and detergent-soluble G protein-coupled receptors in *Escherichia coli*. *Protein Sci.* **17**, 1857–1863.

Liu, Z., Gandhi, C. S., and Rees, D. C. (2009). Structure of a tetrameric MscL in an expanded intermediate state. *Nature* **461**, 120–124.

Locher, K. P., Lee, A. T., and Rees, D. C. (2002). The *E. coli* BtuCD structure: A framework for ABC transporter architecture and mechanism. *Science* **296**, 1091–1098.

Long, S. B., Campbell, E. B., and Mackinnon, R. (2005). Crystal structure of a mammalian voltage-dependent Shaker family K^+ channel. *Science* **309**, 897–903.

Long, S. B., Tao, X., Campbell, E. B., and MacKinnon, R. (2007). Atomic structure of a voltage-dependent K^+ channel in a lipid membrane-like environment. *Nature* **450**, 376–382.

Lundback, A. K., van den Berg, S., Hebert, H., Berglund, H., and Eshaghi, S. (2008). Exploring the activity of tobacco etch virus protease in detergent solutions. *Anal. Biochem.* **382**, 69–71.

Lundstrom, K. (2005). Structural genomics of GPCRs. *Trends Biotechnol.* **23**, 103–108.

Lundstrom, K., Wagner, R., Reinhart, C., Desmyter, A., Cherouati, N., Magnin, T., Zeder-Lutz, G., Courtot, M., Prual, C., Andre, N., Hassaine, G., Michel, H., Cambillau, C., and Pattus, F. (2006). Structural genomics on membrane proteins: Comparison of more than 100 GPCRs in 3 expression systems. *J. Struct. Funct. Genomics* **7**, 77–91.

Lunin, V. V., Dobrovetsky, E., Khutoreskaya, G., Zhang, R., Joachimiak, A., Doyle, D. A., Bochkarev, A., Maguire, M. E., Edwards, A. M., and Koth, C. M. (2006). Crystal structure of the CorA Mg^{2+} transporter. *Nature* **440**, 833–837.

Maeda, S., Nakagawa, S., Suga, M., Yamashita, E., Oshima, A., Fujiyoshi, Y., and Tsukihara, T. (2009). Structure of the connexin 26 gap junction channel at 3.5 Å resolution. *Nature* **458**, 597–602.

Magnani, F., Shibata, Y., Serrano-Vega, M. J., and Tate, C. G. (2008). Co-evolving stability and conformational homogeneity of the human adenosine A2a receptor. *Proc. Natl. Acad. Sci. USA* **105**, 10744–10749.

Makrides, S. C. (1996). Strategies for achieving high-level expression of genes in *Escherichia coli*. *Microbiol. Rev.* **60**, 512–538.

Mancia, F., and Hendrickson, W. A. (2007). Expression of recombinant G-protein coupled receptors for structural biology. *Mol. Biosyst.* **3**, 723–734.

Marsischky, G., and LaBaer, J. (2004). Many paths to many clones: A comparative look at high-throughput cloning methods. *Genome Res.* **14**, 2020–2028.

Martin, A., Damian, M., Laguerre, M., Parello, J., Pucci, B., Serre, L., Mary, S., Marie, J., and Baneres, J. L. (2009). Engineering a G protein-coupled receptor for structural studies: Stabilization of the BLT1 receptor ground state. *Protein Sci.* **18**, 727–734.

Martinac, B., Buechner, M., Delcour, A. H., Adler, J., and Kung, C. (1987). Pressure-sensitive ion channel in *Escherichia coli*. *Proc. Natl. Acad. Sci. USA* **84**, 2297–2301.

Martinez Molina, D., Cornvik, T., Eshaghi, S., Haeggstrom, J. Z., Nordlund, P., and Sabet, M. I. (2008). Engineering membrane protein overproduction in *Escherichia coli*. *Protein Sci.* **17**, 673–680.

Martinez Molina, D., Wetterholm, A., Kohl, A., McCarthy, A. A., Niegowski, D., Ohlson, E., Hammarberg, T., Eshaghi, S., Haeggstrom, J. Z., and Nordlund, P. (2007). Structural basis for synthesis of inflammatory mediators by human leukotriene C4 synthase. *Nature* **448**, 613–616.

Massey-Gendel, E., Zhao, A., Boulting, G., Kim, H. Y., Balamotis, M. A., Seligman, L. M., Nakamoto, R. K., and Bowie, J. U. (2009). Genetic selection system for improving recombinant membrane protein expression in *E. coli*. *Protein Sci.* **18**, 372–383.

Maxfield, F. R., and Mondal, M. (2006). Sterol and lipid trafficking in mammalian cells. *Biochem. Soc. Trans.* **34**, 335–339.

McCusker, E. C., Bane, S. E., O'Malley, M. A., and Robinson, A. S. (2007). Heterologous GPCR expression: A bottleneck to obtaining crystal structures. *Biotechnol. Prog.* **23**, 540–547.

Michalke, K., Graviere, M. E., Huyghe, C., Vincentelli, R., Wagner, R., Pattus, F., Schroeder, K., Oschmann, J., Rudolph, R., Cambillau, C., and Desmyter, A. (2009). Mammalian G-protein-coupled receptor expression in *Escherichia coli*: I. High-throughput large-scale production as inclusion bodies. *Anal. Biochem.* **386**, 147–155.

Midgett, C. R., and Madden, D. R. (2007). Breaking the bottleneck: Eukaryotic membrane protein expression for high-resolution structural studies. *J. Struct. Biol.* **160**, 265–274.

Mio, K., Ogura, T., Yamamoto, T., Hiroaki, Y., Fujiyoshi, Y., Kubo, Y., and Sato, C. (2009). Reconstruction of the P2X(2) receptor reveals a vase-shaped structure with lateral tunnels above the membrane. *Structure* **17**, 266–275.

Miroux, B., and Walker, J. E. (1996). Over-production of proteins in *Escherichia coli*: Mutant hosts that allow synthesis of some membrane proteins and globular proteins at high levels. *J. Mol. Biol.* **260**, 289–298.

Mohanty, A. K., Simmons, C. R., and Wiener, M. C. (2003). Inhibition of tobacco etch virus protease activity by detergents. *Protein Expr. Purif.* **27**, 109–114.

Mohanty, A. K., and Wiener, M. C. (2004). Membrane protein expression and production: Effects of polyhistidine tag length and position. *Protein Expr. Purif.* **33**, 311–325.

Monne, M., Chan, K. W., Slotboom, D. J., and Kunji, E. R. (2005). Functional expression of eukaryotic membrane proteins in *Lactococcus lactis*. *Protein Sci.* **14**, 3048–3056.

Murakami, M., and Kouyama, T. (2008). Crystal structure of squid rhodopsin. *Nature* **453**, 363–367.

Neophytou, I., Harvey, R., Lawrence, J., Marsh, P., Panaretou, B., and Barlow, D. (2007). Eukaryotic integral membrane protein expression utilizing the *Escherichia coli* glycerol-conducting channel protein (GlpF). *Appl. Microbiol. Biotechnol.* **77**, 375–381.

Newby, Z. E., O'Connell, J., 3rd, Robles-Colmenares, Y., Khademi, S., Miercke, L. J., and Stroud, R. M. (2008). Crystal structure of the aquaglyceroporin PfAQP from the malarial parasite *Plasmodium falciparum*. *Nat. Struct. Mol. Biol.* **15**, 619–625.

Newby, Z. E., O'Connell, J. D., 3rd, Gruswitz, F., Hays, F. A., Harries, W. E., Harwood, I. M., Ho, J. D., Lee, J. K., Savage, D. F., Miercke, L. J., and Stroud, R. M. (2009). A general protocol for the crystallization of membrane proteins for X-ray structural investigation. *Nat. Protoc.* **4**, 619–637.

Newstead, S., Kim, H., von Heijne, G., Iwata, S., and Drew, D. (2007). High-throughput fluorescent-based optimization of eukaryotic membrane protein overexpression and purification in *Saccharomyces cerevisiae*. *Proc. Natl. Acad. Sci. USA* **104**, 13936–13941.

Nunez, M. T., and Glass, J. (1982). Reconstitution of the transferrin receptor in lipid vesicles. Effect of cholesterol on the binding of transferrin. *Biochemistry* **21**, 4139–4143.

Okada, T., and Palczewski, K. (2001). Crystal structure of rhodopsin: Implications for vision and beyond. *Curr. Opin. Struct. Biol.* **11**, 420–426.

Opekarova, M., and Tanner, W. (2003). Specific lipid requirements of membrane proteins – A putative bottleneck in heterologous expression. *Biochim. Biophys. Acta* **1610**, 11–22.

Oxenoid, K., and Chou, J. J. (2005). The structure of phospholamban pentamer reveals a channel-like architecture in membranes. *Proc. Natl. Acad. Sci. USA* **102**, 10870–10875.

Page, R. C., Lee, S., Moore, J. D., Opella, S. J., and Cross, T. A. (2009). Backbone structure of a small helical integral membrane protein: A unique structural characterization. *Protein Sci.* **18**, 134–146.

Palanivelu, D. V., Kozono, D. E., Engel, A., Suda, K., Lustig, A., Agre, P., and Schirmer, T. (2006). Co-axial association of recombinant eye lens aquaporin-0 observed in loosely packed 3D crystals. *J. Mol. Biol.* **355**, 605–611.

Palczewski, K. (2006). G protein-coupled receptor rhodopsin. *Annu. Rev. Biochem.* **75**, 743–767.

Palczewski, K., Kumasaka, T., Hori, T., Behnke, C. A., Motoshima, H., Fox, B. A., Le Trong, I., Teller, D. C., Okada, T., Stenkamp, R. E., Yamamoto, M., and Miyano, M. (2000). Crystal structure of rhodopsin: A G protein-coupled receptor. *Science* **289**, 739–745.

Park, J. H., Scheerer, P., Hofmann, K. P., Choe, H. W., and Ernst, O. P. (2008). Crystal structure of the ligand-free G-protein-coupled receptor opsin. *Nature* **454**, 183–187.

Pedersen, B. P., Buch-Pedersen, M. J., Morth, J. P., Palmgren, M. G., and Nissen, P. (2007). Crystal structure of the plasma membrane proton pump. *Nature* **450**, 1111–1114.

Pierce, K. L., Premont, R. T., and Lefkowitz, R. J. (2002). Seven-transmembrane receptors. *Nat. Rev. Mol. Cell. Biol.* **3**, 639–650.

Pinkett, H. W., Lee, A. T., Lum, P., Locher, K. P., and Rees, D. C. (2007). An inward-facing conformation of a putative metal-chelate-type ABC transporter. *Science* **315**, 373–377.

Prilusky, J., and Bibi, E. (2009). Studying membrane proteins through the eyes of the genetic code revealed a strong uracil bias in their coding mRNAs. *Proc. Natl. Acad. Sci. USA* **106**, 6662–6666.

Psakis, G., Nitschkowski, S., Holz, C., Kress, D., Maestre-Reyna, M., Polaczek, J., Illing, G., and Essen, L. O. (2007). Expression screening of integral membrane proteins from *Helicobacter pylori* 26695. *Protein Sci.* **16**, 2667–2676.

Psakis, G., Polaczek, J., and Essen, L. O. (2009). AcrB *et al.*: Obstinate contaminants in a picogram scale. One more bottleneck in the membrane protein structure pipeline. *J. Struct. Biol.* **166**, 107–111.

Qin, H., Hu, J., Hua, Y., Challa, S. V., Cross, T. A., and Gao, F. P. (2008). Construction of a series of vectors for high throughput cloning and expression screening of membrane proteins from Mycobacterium tuberculosis. *BMC Biotechnol.* **8**, 51.

Rasmussen, S. G., Choi, H. J., Rosenbaum, D. M., Kobilka, T. S., Thian, F. S., Edwards, P. C., Burghammer, M., Ratnala, V. R., Sanishvili, R., Fischetti, R. F., Schertler, G. F., Weis, W. I., and Kobilka, B. K. (2007). Crystal structure of the human beta2 adrenergic G-protein-coupled receptor. *Nature* **450**, 383–387.

Ren, H., Yu, D., Ge, B., Cook, B., Xu, Z., and Zhang, S. (2009). High-level production, solubilization and purification of synthetic human GPCR chemokine receptors CCR5, CCR3, CXCR4 and CX3CR1. *PLoS ONE* **4**, e4509.

Rice, A. E., Mendez, M. J., Hokanson, C. A., Rees, D. C., and Bjorkman, P. J. (2009). Investigation of the biophysical and cell biological properties of ferroportin, a multipass integral membrane protein iron exporter. *J. Mol. Biol.* **386**, 717–732.

Roosild, T. P., Greenwald, J., Vega, M., Castronovo, S., Riek, R., and Choe, S. (2005). NMR structure of Mistic, a membrane-integrating protein for membrane protein expression. *Science* **307**, 1317–1321.

Rosenbaum, D. M., Cherezov, V., Hanson, M. A., Rasmussen, S. G., Thian, F. S., Kobilka, T. S., Choi, H. J., Yao, X. J., Weis, W. I., Stevens, R. C., and Kobilka, B. K. (2007). GPCR engineering yields high-resolution structural insights into beta2-adrenergic receptor function. *Science* **318**, 1266–1273.

Roth, C. B., Hanson, M. A., and Stevens, R. C. (2008). Stabilization of the human beta2-adrenergic receptor TM4-TM3-TM5 helix interface by mutagenesis of Glu122(3.41), a critical residue in GPCR structure. *J. Mol. Biol.* **376**, 1305–1319.

Sakai, H., and Tsukihara, T. (1998). Structures of membrane proteins determined at atomic resolution. *J. Biochem.* **124**, 1051–1059.

Sarkar, C. A., Dodevski, I., Kenig, M., Dudli, S., Mohr, A., Hermans, E., and Pluckthun, A. (2008). Directed evolution of a G protein-coupled receptor for expression, stability, and binding selectivity. *Proc. Natl. Acad. Sci. USA* **105**, 14808–14813.

Savage, D. F., Anderson, C. L., Robles-Colmenares, Y., Newby, Z. E., and Stroud, R. M. (2007). Cell-free complements *in vivo* expression of the *E. coli* membrane proteome. *Protein Sci.* **16**, 966–976.

Scheerer, P., Park, J. H., Hildebrand, P. W., Kim, Y. J., Krauss, N., Choe, H. W., Hofmann, K. P., and Ernst, O. P. (2008). Crystal structure of opsin in its G-protein-interacting conformation. *Nature* **455**, 497–502.

Schnell, J. R., and Chou, J. J. (2008). Structure and mechanism of the M2 proton channel of influenza A virus. *Nature* **451**, 591–595.

Schwarz, D., Dotsch, V., and Bernhard, F. (2008). Production of membrane proteins using cell-free expression systems. *Proteomics* **8**, 3933–3946.

Schwarz, D., Junge, F., Durst, F., Frolich, N., Schneider, B., Reckel, S., Sobhanifar, S., Dotsch, V., and Bernhard, F. (2007). Preparative scale expression of membrane proteins in *Escherichia coli*-based continuous exchange cell-free systems. *Nat. Protoc.* **2**, 2945–2957.

Screpanti, E., Padan, E., Rimon, A., Michel, H., and Hunte, C. (2006). Crucial steps in the structure determination of the Na^+/H^+ antiporter NhaA in its native conformation. *J. Mol. Biol.* **362**, 192–202.

Seidel, H. M., Pompliano, D. L., and Knowles, J. R. (1992). Phosphonate biosynthesis: Molecular cloning of the gene for phosphoenolpyruvate mutase from *Tetrahymena pyriformis* and overexpression of the gene product in *Escherichia coli*. *Biochemistry* **31**, 2598–2608.

Serrano-Vega, M. J., Magnani, F., Shibata, Y., and Tate, C. G. (2008). Conformational thermostabilization of the beta1-adrenergic receptor in a detergent-resistant form. *Proc. Natl. Acad. Sci. USA* **105**, 877–882.

Shi, N., Ye, S., Alam, A., Chen, L., and Jiang, Y. (2006). Atomic structure of a Na^+- and K^+-conducting channel. *Nature* **440**, 570–574.

Singh, V. (1999). Disposable bioreactor for cell culture using wave-induced agitation. *Cytotechnology* **30**, 149–158.

Skretas, G., and Georgiou, G. (2008). Engineering G protein-coupled receptor expression in bacteria. *Proc. Natl. Acad. Sci. USA* **105**, 14747–14748.

Sorensen, H. P., and Mortensen, K. K. (2005). Advanced genetic strategies for recombinant protein expression in *Escherichia coli*. *J. Biotechnol.* **115**, 113–128.

Stanasila, L., Massotte, D., Kieffer, B. L., and Pattus, F. (1999). Expression of delta, kappa and mu human opioid receptors in *Escherichia coli* and reconstitution of the high-affinity state for agonist with heterotrimeric G proteins. *Eur. J. Biochem.* **260**, 430–438.

Standfuss, J., Xie, G., Edwards, P. C., Burghammer, M., Oprian, D. D., and Schertler, G. F. (2007). Crystal structure of a thermally stable rhodopsin mutant. *J. Mol. Biol.* **372**, 1179–1188.

Stein, A., Weber, G., Wahl, M. C., and Jahn, R. (2009). Helical extension of the neuronal SNARE complex into the membrane. *Nature* **460**, 525–528.

Stroud, R. M., Choe, S., Holton, J., Kaback, H. R., Kwiatkowski, W., Minor, D. L., Riek, R., Sali, A., Stahlberg, H., and Harries, W. (2009). 2007 annual progress report synopsis of the Center for Structures of Membrane Proteins. *J. Struct. Funct. Genomics* **10**, 193–208.

Studier, F. W. (2005). Protein production by auto-induction in high density shaking cultures. *Protein Expr. Purif.* **41**, 207–234.

Studier, F. W., Rosenberg, A. H., Dunn, J. J., and Dubendorff, J. W. (1990). Use of T7 RNA polymerase to direct expression of cloned genes. *Methods Enzymol.* **185**, 60–89.

Sugawara, T., Ito, K., Shiroishi, M., Tokuda, N., Asada, H., Yurugi-Kobayashi, T., Shimamura, T., Misaka, T., Nomura, N., Murata, T., Abe, K., Iwata, S., and Kobayashi, T. (2009). Fluorescence-based optimization of human bitter taste receptor expression in *Saccharomyces cerevisiae*. *Biochem. Biophys. Res. Commun.* **382**, 704–710.

Surade, S., Klein, M., Stolt-Bergner, P. C., Muenke, C., Roy, A., and Michel, H. (2006). Comparative analysis and "expression space" coverage of the production of prokaryotic membrane proteins for structural genomics. *Protein Sci.* **15**, 2178–2189.

Tani, K., Mitsuma, T., Hiroaki, Y., Kamegawa, A., Nishikawa, K., Tanimura, Y., and Fujiyoshi, Y. (2009). Mechanism of aquaporin-4's fast and highly selective water conduction and proton exclusion. *J. Mol. Biol.* **389**, 694–706.

Tao, Y. X. (2008). Constitutive activation of G protein-coupled receptors and diseases: Insights into mechanisms of activation and therapeutics. *Pharmacol Ther.* **120**, 129–148.

Tate, C. G., Haase, J., Baker, C., Boorsma, M., Magnani, F., Vallis, Y., and Williams, D. C. (2003). Comparison of seven different heterologous protein expression systems for the production of the serotonin transporter. *Biochim. Biophys. Acta* **1610**, 141–153.

Ton, V. K., and Rao, R. (2004). Functional expression of heterologous proteins in yeast: Insights into Ca^{2+} signaling and Ca^{2+}-transporting ATPases. *Am. J. Physiol. Cell Physiol.* **287**, C580–C589.

Tornroth-Horsefield, S., Wang, Y., Hedfalk, K., Johanson, U., Karlsson, M., Tajkhorshid, E., Neutze, R., and Kjellbom, P. (2006). Structural mechanism of plant aquaporin gating. *Nature* **439**, 688–694.

Tucker, J., and Grisshammer, R. (1996). Purification of a rat neurotensin receptor expressed in *Escherichia coli*. *Biochem. J.* **317**(Pt 3), 891–899.

Urban, S., and Wolfe, M. S. (2005). Reconstitution of intramembrane proteolysis in vitro reveals that pure rhomboid is sufficient for catalysis and specificity. *Proc. Natl. Acad. Sci. USA* **102**, 1883–1888.

van den Berg, B., Clemons, W. M., Jr., Collinson, I., Modis, Y., Hartmann, E., Harrison, S. C., and Rapoport, T. A. (2004). X-ray structure of a protein-conducting channel. *Nature* **427**, 36–44.

Veesler, D., Blangy, S., Cambillau, C., and Sciara, G. (2008). There is a baby in the bath water: AcrB contamination is a major problem in membrane-protein crystallization. *Acta Crystallogr. Sect. F Struct. Biol. Cryst. Commun.* **64**, 880–885.

Wagner, S., Baars, L., Ytterberg, A. J., Klussmeier, A., Wagner, C. S., Nord, O., Nygren, P. A., van Wijk, K. J., de Gier, J. W. (2007). Consequences of membrane protein overexpression in *Escherichia coli*. *Mol Cell Proteomics.* **6**, 1527–1550.

Wagner, S., Bader, M. L., Drew, D., de Gier, J. W. (2006). Rationalizing membrane protein overexpression. *Trends Biotechnol.* **24**, 364–371.

Wagner, S., Klepsch, M. M., Schlegel, S., Appel, A., Draheim, R., Tarry, M., Hogbom, M., van Wijk, K. J., Slotboom, D. J., Persson, J. O., de Gier, J. W. (2008). Tuning *Escherichia coli* for membrane protein overexpression. *Proc. Natl. Acad. Sci. USA* **105**, 14371–14376.

Walhout, A. J., Temple, G. F., Brasch, M. A., Hartley, J. L., Lorson, M. A., van den Heuvel, S., and Vidal, M. (2000). GATEWAY recombinational cloning: Application to the cloning of large numbers of open reading frames or ORFeomes. *Methods Enzymol.* **328**, 575–592.

Wang, D. N., Safferling, M., Lemieux, M. J., Griffith, H., Chen, Y., and Li, X. D. (2003). Practical aspects of overexpressing bacterial secondary membrane transporters for structural studies. *Biochim. Biophys. Acta.* **1610**, 23–36.

Wang, W., Black, S. S., Edwards, M. D., Miller, S., Morrison, E. L., Bartlett, W., Dong, C., Naismith, J. H., and Booth, I. R. (2008). The structure of an open form of an *E. coli* mechanosensitive channel at 3.45 Å resolution. *Science* **321**, 1179–1183.

Warne, T., Serrano-Vega, M. J., Baker, J. G., Moukhametzianov, R., Edwards, P. C., Henderson, R., Leslie, A. G., Tate, C. G., and Schertler, G. F. (2008). Structure of a beta1-adrenergic G-protein-coupled receptor. *Nature* **454**, 486–491.

Warne, T., Serrano-Vega, M. J., Tate, C. G., and Schertler, G. F. (2009). Development and crystallization of a minimal thermostabilised G protein-coupled receptor. *Protein Expr. Purif.* **65**, 204–213.

Waugh, D. S. (2005). Making the most of affinity tags. *Trends Biotechnol.* **23**, 316–320.

Wedekind, A., O'Malley, M. A., Niebauer, R. T., and Robinson, A. S. (2006). Optimization of the human adenosine A2a receptor yields in *Saccharomyces cerevisiae. Biotechnol. Prog.* **22**, 1249–1255.

Weiss, H. M., and Grisshammer, R. (2002). Purification and characterization of the human adenosine A(2a) receptor functionally expressed in *Escherichia coli. Eur. J. Biochem.* **269**, 82–92.

White, M. A., Clark, K. M., Grayhack, E. J., and Dumont, M. E. (2007). Characteristics affecting expression and solubilization of yeast membrane proteins. *J. Mol. Biol.* **365**, 621–636.

Xia, Y., Chhajlani, V., and Wikberg, J. E. (1993). Functional expression of rat alpha 2B-adrenoceptor in *Escherichia coli. Eur. J. Pharmacol.* **246**, 129–133.

Yeliseev, A., Zoubak, L., and Gawrisch, K. (2007). Use of dual affinity tags for expression and purification of functional peripheral cannabinoid receptor. *Protein Expr. Purif.* **53**, 153–163.

Yernool, D., Boudker, O., Jin, Y., and Gouaux, E. (2004). Structure of a glutamate transporter homologue from *Pyrococcus horikoshii. Nature* **431**, 811–818.

Yurugi-Kobayashi, T., Asada, H., Shiroishi, M., Shimamura, T., Funamoto, S., Katsuta, N., Ito, K., Sugawara, T., Tokuda, N., Tsujimoto, H., Murata, T., Nomura, N., Haga, K., Haga, T., Iwata, S., and Kobayashi, T. (2009). Comparison of functional non-glycosylated GPCRs expression in *Pichia pastoris. Biochem. Biophys. Res. Commun.* **380**, 271–276.

Zeder-Lutz, G., Cherouati, N., Reinhart, C., Pattus, F., and Wagner, R. (2006). Dot-blot immunodetection as a versatile and high-throughput assay to evaluate recombinant GPCRs produced in the yeast *Pichia pastoris. Protein Expr. Purif.* **50**, 118–127.

Zhang, G., Hubalewska, M., and Ignatova, Z. (2009). Transient ribosomal attenuation coordinates protein synthesis and co-translational folding. *Nat. Struct. Mol. Biol.* **16**, 274–280.

Zhou, Y., Morais-Cabral, J. H., Kaufman, A., and MacKinnon, R. (2001). Chemistry of ion coordination and hydration revealed by a K^+ channel-Fab complex at 2.0 Å resolution. *Nature* **414**, 43–48.

AUTHOR INDEX

A

B

C

D

E

F

G

H

I

J

N

O

P

Q

R

S

SUBJECT INDEX

Note: The letters 'f' and 't' following the locators refer to figures and tables respectively

A

B

C

D

E

F

G

H

I

K

L

M

N

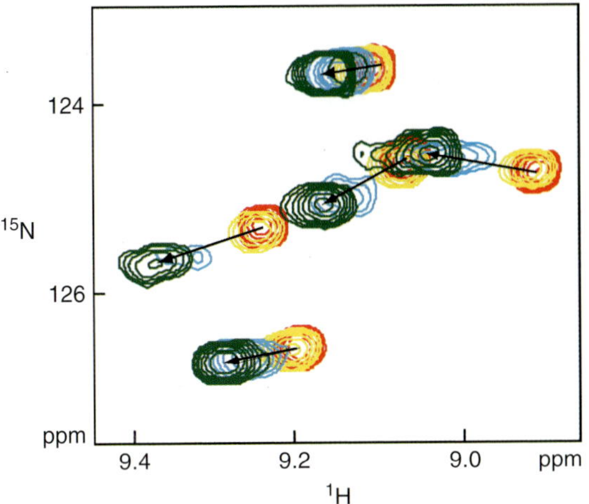

Matte *et al.*, Chapter 1, Fig 2. NMR analysis of Ste50_RA/Opy2 peptide interactions. Overlay of regions of ^{1}H-^{15}N HSQC spectra for solutions containing a free RA domain (red) and increasing ratios of unlabeled Opy2 peptide titrated into ^{15}N-RA domain (orange, blue, and green, respectively). The arrows indicate displacement of selected ^{15}N-RA domain peaks.

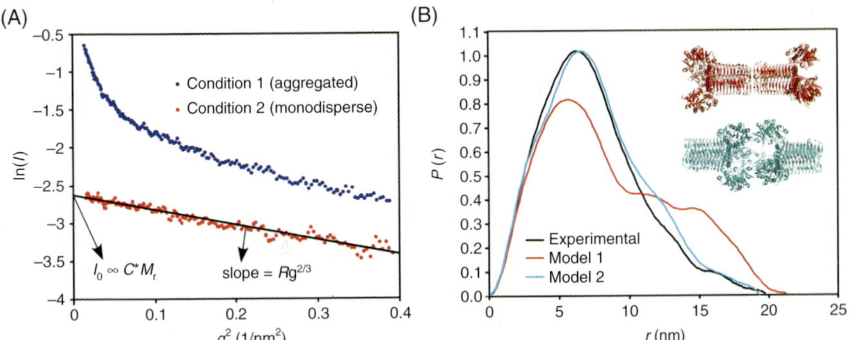

Matte *et al.*, Chapter 1, Fig 3. Use of SAXS in structural genomics of protein complexes. (A) The Guinier plot, which is the square of scattering vector q ($4\pi \sin \theta/\lambda$) versus the natural logarithm of the intensity, can be used as a diagnostic of protein aggregation and size in solution. An example is shown here for a protein complex concentrated to 4 mg/ml in two different buffer conditions. Monodisperse globular particle solution should give a linear Guinier within the range $q_{\min}$ to $q_{\max}*R_g < 1.3$, as observed in condition 2. The slope enables calculation of R_g whereas the intercept enables the molecular weight to be determined if the concentration is known accurately. (B) The pair-density distribution function $P(r)$ allows the identification of a biological unit from a crystal structure that gives two possible hexameric forms (models 1 and 2). The experimental $P(r)$ function is much more similar to the hexamer #2, which implies that the protein forms hexamer of this type in solution.

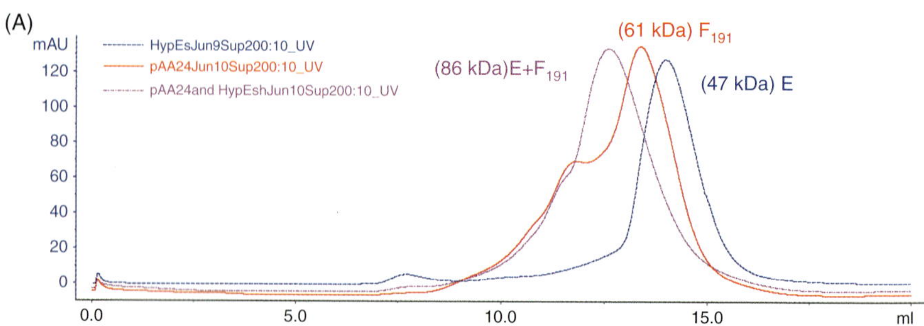

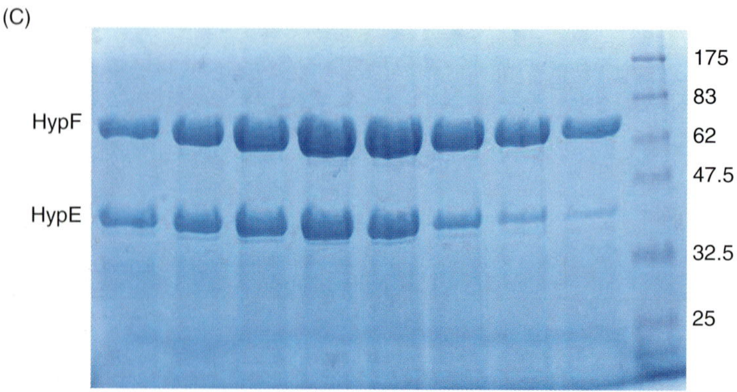

MATTE *ET AL.*, CHAPTER 1, FIG 4. Protein–protein complex between *E. coli* HypE and a deletion construct (Δ1–191) of *E. coli* HypF.